Sequence Data Mining

ADVANCES IN DATABASE SYSTEMS

Series Editor

Ahmed K. Elmagarmid

Purdue University
West Lafayette, IN 47907

Other books in the Series:

DATA STREAMS: ***Models and Algorithms***, *edited by Charu C. Aggarwal*; ISBN: 978- 0-387-28759-1

SIMILARITY SEARCH: ***The Metric Space Approach***, *P. Zezula, G. Amato, V. Dohnal, M. Batko*; ISBN: 0-387-29146-6

STREAM DATA MANAGEMENT, *Nauman Chaudhry, Kevin Shaw, Mahdi Abdelguerfi*; ISBN: 0-387-24393-3

FUZZY DATABASE MODELING WITH XML, *Zongmin Ma;* ISBN: 0-387-24248-1

MINING SEQUENTIAL PATTERNS FROM LARGE DATA SETS, *Wei Wang and Jiong Yang;* ISBN: 0-387-24246-5

ADVANCED SIGNATURE INDEXING FOR MULTIMEDIA AND WEB APPLICATIONS, *Yannis Manolopoulos, Alexandros Nanopoulos, Eleni Tousidou;* ISBN: 1-4020-7425-5

ADVANCES IN DIGITAL GOVERNMENT*: Technology, Human Factors, and Policy*, *edited by William J. McIver, Jr. and Ahmed K. Elmagarmid;* ISBN: 1-4020-7067-5

INFORMATION AND DATABASE QUALITY, *Mario Piattini, Coral Calero and Marcela Genero;* ISBN: 0-7923- 7599-8

DATA QUALITY, *Richard Y. Wang, Mostapha Ziad, Yang W. Lee:* ISBN: 0-7923-7215-8

THE FRACTAL STRUCTURE OF DATA REFERENCE: *Applications to the Memory Hierarchy*, *Bruce McNutt;* ISBN: 0-7923-7945-4

SEMANTIC MODELS FOR MULTIMEDIA DATABASE SEARCHING AND BROWSING, *Shu-Ching Chen, R.L. Kashyap, and Arif Ghafoor*; ISBN: 0-7923-7888-1

INFORMATION BROKERING ACROSS HETEROGENEOUS DIGITAL DATA: A Metadata-based Approach, *Vipul Kashyap, Amit Sheth*; ISBN: 0-7923-7883-0

DATA DISSEMINATION IN WIRELESS COMPUTING ENVIRONMENTS, *Kian-Lee Tan and Beng Chin Ooi*; ISBN: 0-7923-7866-0

MIDDLEWARE NETWORKS: Concept, Design and Deployment of Internet Infrastructure, *Michah Lerner, George Vanecek, Nino Vidovic, Dad Vrsalovic;* ISBN: 0-7923-7840-7

ADVANCED DATABASE INDEXING, *Yannis Manolopoulos, Yannis Theodoridis, Vassilis J. Tsotras;* ISBN: 0-7923-7716-8

MULTILEVEL SECURE TRANSACTION PROCESSING, *Vijay Atluri, Sushil Jajodia, Binto George* ISBN: 0-7923-7702-8

FUZZY LOGIC IN DATA MODELING, *Guoqing Chen* ISBN: 0-7923-8253-6

For a complete listing of books in this series, go to http://www.springer.com

Sequence Data Mining

by

Guozhu Dong
Wright State University
Dayton, Ohio, USA

and

Jian Pei
Simon Fraser University
Burnaby, BC, Canada

Guozhu Dong, PhD, Professor
Department of Computer Science and Eng.
Wright State University
Dayton, Ohio, 45435, USA
e-mail: guozhu.dong@wright.edu

Jian Pei, Ph.D.
Assistant Professor
School of Computing Science
Simon Fraser University
8888 University Drive
Burnaby, BC Canada V5A 1S6
e-mail: jpei@cs.sfu.ca

ISBN-13: 978-1-4419-4352-1 e-ISBN-13: 978-0-387-69937-0

Softcover reprint of the hardcover 1st edition 2007

Printed on acid-free paper.

9 8 7 6 5 4 3 2 1

springer.com

To my parents, my wife and my children. {G.D.}
To my wife Jennifer. {J.P.}

Foreword

With the rapid development of computer and Internet technology, tremendous amounts of data have been collected in various kinds of applications, and *data mining*, *i.e.*, finding interesting patterns and knowledge from a vast amount of data, has become an imminent task. Among all kinds of data, sequence data has its own unique characteristics and importance, and claims many interesting applications. From customer shopping transactions, to global climate change, from web click streams to biological DNA sequences, the sequence data is ubiquitous and poses its own challenging research issues, calling for dedicated treatment and systematic analysis.

Despite of the existence of a lot of general data mining algorithms and methods, sequence data mining deserves dedicated study and in-depth treatment because of its unique nature of ordering, which leads to many interesting new kinds of knowledge to be discovered, including sequential patterns, motifs, periodic patterns, partially ordered patterns, approximate biological sequence patterns, and so on; and these kinds of patterns will naturally promote the development of new classification, clustering and outlier analysis methods, which in turn call for new, diverse application developments. Therefore, *sequence data mining*, *i.e.*, mining patterns and knowledge from large amount of sequence data, has become one of the most essential and active subfields of data mining research. With many years of active research on sequence data mining by data mining, machine learning, statistical data analysis, and bioinformatics researchers, it is time to present a systematic introduction and comprehensive overview of the state-of-the-art of this interesting theme. This book, by Professors Guozhu Dong and Jian Pei, serves this purpose timely, with remarkable conciseness and in great quality.

There have been many books on the general principles and methodologies of data mining. However, the diversities of data and applications call for dedicated, in-depth, and thorough treatment of each specific kind of data, and for each kind of data, compile a vast array of techniques from multiple disciplines into one comprehensive but concise introduction. Thus there is no wonder to see the recent trend of the publication of a series of new, domain-specific data mining books, such as those on Web data mining, stream data mining, geo-spatial data mining, and multimedia data mining. This book integrates the methodologies of sequence data mining developed in multiple disciplines, including data mining, machine learning, statistics, bioinformatics, genomics, web services, and financial data analysis, into one comprehensive and easily-accessible introduction. It starts with a general overview of the sequence data mining problem, by characterizing the sequence data, sequence patterns and sequence models and their various applications, and then proceeds to different mining algorithms and methodologies. It covers a set of exciting research themes, including sequential pattern mining methods; classification, clustering and feature extraction of sequence data; identification and characterization of sequence motifs; mining partial orders from sequences; distinguishing sequence patterns; and other interesting related topics. The scope of the book is broad, nevertheless the treatment of each chapter is rigorous, in sufficient depth, but still easy to read and comprehend.

Both authors of the book are prominent researchers on sequence data mining and have made important contributions to the progress of this dynamic research field. This ensures that the book is authoritative and reflects the current state of the art. Nevertheless, the book gives a balanced treatment on a wide spectrum of topics, far beyond the authors' own methodologies and research scopes.

Sequence data mining is still a fairly young and dynamic research field. This book may serve researcher and application developers a comprehensive overview of the general concepts, techniques, and applications on sequence data mining and help them explore this exciting field and develop new methods and applications. It may also serve graduate students and other interested readers a general introduction to the state-of-the-art of this promising field.

I find the book is enjoyable to read. I hope you like it too.

Jiawei Han
University of Illinois, Urbana-Champaign
April 29, 2007

Biography

Jiawei Han, University of Illinois at Urbana-Champaign

Jiawei Han, Professor, Department of Computer Science, University of Illinois at Urbana-Champaign. His research includes data mining, data warehousing, database systems, data mining from spatiotemporal data, multimedia data, stream and RFID data, Web data, social network data, and biological data, with over 300 journal and conference publications. He has chaired or served on over 100 program committees of international conferences and workshops, including PC co-chair of 2005 (IEEE) International Conference on Data Mining (ICDM).

He is an ACM Fellow and has received 2004 ACM SIGKDD Innovations Award and 2005 IEEE Computer Society Technical Achievement Award. His book "Data Mining: Concepts and Techniques" (2nd ed., Morgan Kaufmann, 2006) has been popularly used as a textbook worldwide.

Preface

Sequence data is pervasive in our lives. For example, your schedule for any given day is a sequence of your activities. When you read a news story, you are told the development of some events which is also a sequence. If you have investment in companies, you are keen to study the history of those companies' stocks. Deep in your life, you rely on biological sequences including DNA and RNA sequences.

Understanding sequence data is of grand importance. As early as our history can call, our ancestors already started to make predictions or simply conjectures based on their observations of event sequences. For example, a typical task of royal astronomers in ancient China was to make conjectures according to their observations of stellar movements. Even much earlier before that, the nature encodes some "sequence learning algorithms" in lives. For example, some animals such as dogs, mice, and snakes have the capability to predict earthquakes based on environmental change sequences, though the mechanisms are still largely mysteries.

When the general field of data mining emerged in the 1990s, sequence data mining naturally became one of the first class citizens in the field. Much research has been conducted on sequence data mining in the last dozen years. Hundreds if not thousands of research papers have been published in forums of various disciplines, such as data mining, database systems, information retrieval, biology and bioinformatics, industrial engineering, etc. The area of sequence data mining has developed rapidly, producing a diversified array of concepts, techniques and algorithmic tools.

The purpose of this book is to provide, in one place, a concise introduction to the field of sequence data mining, and a fairly comprehensive overview of the essential research results. After an introduction to the basics of sequence data mining, the major topics include (1) mining frequent and closed sequential patterns, (2) clustering, classification, features and distances of sequence data, (3) sequence motifs – identifying and characterizing sequence families, (4) mining partial orders from sequences, (5) mining distinguishing sequence patterns, and (6) overviewing some related topics.

This monograph can be useful to academic researchers and graduate students interested in data mining in general and in sequence data mining in particular, and to scientists and engineers working in fields where sequence data mining is involved, such as bioinformatics, genomics, web services, security, and financial data analysis.

Although sequence data mining is discussed in some general data mining textbooks, as you will see in your reading of our book, we conduct a much deeper and more thorough treatment of sequence data mining, and we draw connections to applications whenever it is possible. Therefore, this manuscript covers much more on sequence data mining than a general data mining textbook.

The area of sequence data mining, although a sub-field of general data mining, is now very rich and it is impossible to cover all of its aspects in this book. Instead, in this book, we tried our best to select several important and fundamental topics, and to provide introductions to the essential concepts and methods, of this rich area.

Sequence data mining is still a fairly young research field. Much more remains to be discovered in this exciting research direction, regarding general concepts, techniques, and applications. We invite you to enjoy the exciting exploration.

Acknowledgement

Writing a monograph is never easy. We are sincerely grateful to Jiawei Han for his consistent encouragement since the planning stage for this book, as well as writing the foreword for the book. Our deep gratitude also goes to Limsoon Wong and James Bailey for providing very helpful comments on the book. We thank Bin Zhou and Ming Hua for their help in proofreading the draft of this book.

Guozhu Dong is also grateful to Limsoon Wong for introducing him to bioinformatics in the late 1990s. Part of this book was planned and written while he was on sabbatical between 2005 and 2006; he wishes to thank his hosts during this period.

Jian Pei is deeply grateful to Jiawei Han as a mentor for continuous encouragement and support. Jian Pei also thanks his collaborators in the past who have fun together in solving data mining puzzles.

Guozhu Dong
Wright State University
Jian Pei
Simon Fraser University
April, 2007

Contents

1

Introduction

Sequences are an important type of data which occur frequently in many scientific, medical, security, business and other applications. For example, DNA sequences encode the genetic makeup of humans and all species, and protein sequences describe the amino acid composition of proteins and encode the structure and function of proteins. Moreover, sequences can be used to capture how individual humans behave through various temporal activity histories such as weblogs and customer purchase histories. Sequences can also be used to describe how organizations behave through sales histories such as the total sales of various items over time for a supermarket, etc.

Huge amounts of sequence data have been and continue to be collected in genomic and medical studies, in security applications, in business applications, etc. In these applications, the analysis of the data needs to be carried out in different ways to satisfy different application requirements, and it needs to be carried out in an efficient manner. Sequence data mining provides the necessary tools and approaches for unlocking useful knowledge hidden in the mountains of sequence data. The purpose of this book is to present some of the main concepts, techniques, algorithms, and references on sequence data mining.

This introductory chapter has four goals. First, it will provide some example applications of sequence data. Second, it will define several basic/generic concepts for sequences and sequence data mining. Third, it will discuss the major issues of interest in data mining research. Fourth, it will give an overview of the entire book.

1.1 Examples and Applications of Sequence Data

This section describes typical applications and common types of sequence data. It will demonstrate the richness of the types of sequence data, and serve as illustration of some formal concepts to be given in the next section.

1.1.1 Examples of Sequence Data

Biological Sequences: DNA, RNA and Protein

Biological sequences are useful for understanding the structures and functions of various molecules, and for diagnosing and treating diseases. Three major types of biological sequences are deoxyribonucleic acid (DNA) sequences, amino acid (also called peptide or protein) sequences, and ribonucleic acid (RNA) sequences. Figures 1.1 and 1.2 show respectively a part of a DNA sequence and a part of a protein sequence. RNA sequences are slightly different from DNA sequences. Below we briefly discuss some background information on these biological sequences.

The complete set of instructions for making an organism is called the organism's genome. A genome is often encoded in the DNA, which is a long polymer[1] made from four types of nucleotides: adenine (abbreviated as A), cytosine (abbreviated as C), guanine (abbreviated as G) and thymine (abbreviated as T). The DNA contains both the genes, which encode the sequences of proteins, and the non-coding sequences.

GAATTCTCTGTAACACTAAGCTCTCTTCCTCAAAACCAGAGGTAGATAGA
ATGTGTAATAATTTACAGAATTTCTAGACTTCAACGATCTGATTTTTTAA
ATTTATTTTTATTTTTTCAGGTTGAGACTGAGCTAAAGTTAATCTGTGGC

Fig. 1.1. A DNA sequence fragment.

Proteins are polymers made from 20 different amino acids, using information present in genes. Genes are transcribed into RNA; RNA is then subject to post-transcriptional modification and control, resulting in a mature messenger RNA (mRNA); the mRNA is translated by ribosomes into the amino acids of the corresponding proteins. Each amino acid is the translation of a sequence interval of length 3 in the mRNA, which is also called a codon. The 20 amino acids are abbreviated as A, C, D, E, F, G, H, I, K, L, M, N, P, Q, R, S, T, V, W, and Y, respectively. RNA is made from four types of nucleotides: adenine (A), guanine (G), cytosine (C), and uracil (U). The first three are the same as those found in DNA, and uracil replaces thymine as the base complementary to adenine.

There are many data analysis problems of biological interest. Some examples include

- identifying genes and gene start sites from DNA sequences;
- identifying intron/exon splice sites from DNA sequences;
- identifying transcription promotors etc from DNA sequences;

[1] A polymer is a generic term referring to a very long molecule consisting of structural units and repeating units connected by covalent chemical bonds.

SSQIRQNYSTEVEAAVNRLVNLYLRASYTYLSLGFYFDRDDVALEGVCHFF
RELAEEKREGAERLLKMQNQRGGRALFQDLQKPSQDEWGTTPDAMKAA
IVLEKSLNQALLDLHALGSAQADPHLCDFLESHFLDEEVKLIKKMGDHLTN
IQRLVGSQAGLGEYLFERLTLKHD

Fig. 1.2. A protein sequence fragment.

- identifying non-coding RNA (also called small RNA) etc from RNA sequences;
- analyzing the structure and function of proteins from protein sequences;
- identifying the characteristic (motif) patterns of families of DNA, RNA or protein sequences;
- identifying useful sequence families; and
- comparing sequence families (e.g. comparing families associated with different species/diseases).

Advances on these problems can help us to better understand life and diseases.

Event Sequences: Weblogs, System Traces, Purchase Histories and Sales Histories

A major category of sequences are event sequences. Such sequences can be used to understand how the underlying actors (namely the objects which generated the event sequences) of the event sequences behave and how to best deal with them. The following are examples of event sequences.

A weblog is a sequence of user-identifier and event pairs (and perhaps other relevant information). An event is a request of some web resource such as a page (usually identified by the URL of the page) or a service. For each page requested, some additional information may be available, such as the type and the content of the page, and the amount of time the user spent on the page. The events in a weblog are listed in the timestamp ascending order. Figure 1.3 shows an example weblog, where a, b, c, d, e are events, and 100, 200, 300, and 400 are user identifiers. A weblog can also be restricted to a single user.

$$\langle 100, a\rangle, \langle 100, b\rangle, \langle 200, a\rangle, \langle 300, b\rangle, \langle 200, b\rangle, \langle 400, a\rangle, \langle 100, a\rangle, \langle 400, b\rangle,$$
$$\langle 300, a\rangle, \langle 100, c\rangle, \langle 200, c\rangle, \langle 400, a\rangle, \langle 400, e\rangle$$

Fig. 1.3. A weblog sequence.

System traces are similar to weblogs in form. They are sequences of records concerning operations performed by various users/processes to various data and resources in one or more systems.

Customer purchase histories are sequences of tuples, each consisting of a customer identifier, a location, a time, and a set of items purchased, etc. Figure 1.4 shows an example.

$$\langle 223100, 05/26/06, 10am, CentralStation, \{WholeMealBread, AppleJuice\}\rangle,$$
$$\langle 225101, 05/26/06, 11am, CentralStation, \{Burger, Pepsi, Banana\}\rangle,$$
$$\langle 223100, 05/26/06, 4pm, WalMart, \{Milk, Cereal, Vegetable\}\rangle,$$
$$\langle 223100, 05/27/06, 10am, CentralStation, \{WholeMealBread, AppleJuice\}\rangle,$$
$$\langle 225101, 05/27/06, 12noon, CentralStation, \{Burger, Coke, Apple\}\rangle$$

Fig. 1.4. A customer purchase history.

Storewide sales histories are sequences of tuples, each consisting of a store ID, a time (period), the total sales of individual items for the time (period), and other relevant information. Such histories can also contain customer group information and some other information for the sales. Figure 1.5 shows an example.

$$\langle 97100, 05/06, \{\langle Apple : \$85K\rangle, \langle Bread : \$100K\rangle, \langle Cereal : \$150K\rangle, ...\}\rangle,$$
$$\langle 90089, 05/06, \{\langle Apple : \$65K\rangle, \langle Bread : \$105K\rangle, \langle Diaper : \$20K\rangle, ...\}\rangle,$$
$$\langle 97100, 06/06, \{\langle Apple : \$95K\rangle, \langle Bread : \$110K\rangle, \langle Cereal : \$160K\rangle, ...\}\rangle,$$
$$\langle 90089, 06/06, \{\langle Apple : \$66K\rangle, \langle Bread : \$95K\rangle, \langle Diaper : \$22K\rangle, ...\}\rangle$$

Fig. 1.5. A storewide sales history.

1.1.2 Examples of Sequence Mining Applications

We now discuss some example data mining applications on event sequences.

Mining Frequent Subsequences

Ada is a marketing manager in a store. She wants to design a marketing campaign which consists of two major aspects. First, a set of products should be identified for promotion. Hopefully, for promoting those products, customers will be retained, and sales on other products will be stimulated. Second, a set of customers should be targeted so that the promotion information should be delivered.

To start with, Ada has the transactions of customers in the past. Each transaction includes the customer-id, the products bought in the transaction, and the timestamp of the transaction. Grouping transactions by customers and sorting them in the timestamp ascending order, Ada can get a purchase sequence database where each sequence records the behavior of a customer.

Ada may want to find frequent subsequences that are shared by many customers. As patterns, those frequent subsequences can help her to understand the behavior of customers. She can also identify products to be promoted according to the purchase patterns, and the target customers.

Classification of Sequences

Bob is a safety manager in an airline in charge of braking systems in airplanes. A sequence of status records is maintained for each aircraft. Maintaining the braking system of an airplane in a hub airport of the airline is highly desirable since maintenance cost is often several times higher when the job is done in a guest airport. On the other hand, being too proactive in maintenance may also lead to unnecessary cost since parts may be replaced too early and are not fully used.

Therefore, Bob is facing such a question: given an airplane's sequence of status records, predict in high confidence whether the plane needs a maintenance before it goes to the next hub airport. This is a classification problem (or as known as supervised learning) since the prediction is made based on some historical data, that is, some records of previous maintenances collected for references.

Clustering of Sequences

Carol is a medical analyst in charge of analyzing patients' reactions to a new drug. For each patient taking the drug (which is referred to as a case), she collects the sequence of reactions of the patient such as the changes in temperature, blood pressure, and so on. Typically, there are a good number, from 20 to more than 100, of such test cases. In order to summarize the results, she needs to categorize the cases into a few groups – all cases in a group are similar to each other, and the cases in different groups are substantially different from each other.

This is a clustering task (or as known as unsupervised learning), since the sequences are not labeled and the groups should be defined by Carol based on the similarity among sequences.

Other Examples

It is easy to name another dozens of examples of sequence data mining. For example, by mining music sequences, we can predict the composers of music pieces. As another example, an interactive computer game can learn from players' behavior sequences to make it more intelligent and more fun.

The point we want to illustrate here is that sequence data mining is very practical in our lives, which makes it attractive for many researchers and developers.

1.2 Basic Definitions

This section defines the concepts of sequences, sequence types, sequence patterns[2], sequence models, pattern matching, and support; it also discusses major characteristics of sequence data. Some of the definitions are generic, because there is considerable variation between specific instances in different applications. Examples of sequences were given in the previous section, and examples of the other concepts will be given in later chapters.

1.2.1 Sequences and Sequence Types

There is a rich variety of sequence types, ranging from simple sequences of letters to complex sequences of relations. Here we provide a very general definition which can capture most practical examples.

For a given application, sequences are constructed from members of some appropriate element types.

Definition 1.1. Element types *are data types constructed from simple data types using several constructs; some common examples are the following:*

- *An* item type *is a finite set Σ of distinct items. Each $x \in \Sigma$ is a member of the type. For example, the DNA sequences are constructed from the item type of $\Sigma = \{A, C, G, T\}$. We will frequently refer to the items as letters or symbols.*
- *A* set type *has the form 2^τ, where τ is an element type. A member of this type is a finite set of members of type τ.*
  In particular, for each finite set Σ of distinct items, 2^Σ is a set type commonly referred to as a basket type. *For example, market basket sequences are constructed from the element type of 2^Σ, where Σ is a fixed set of items.*
- *A* tuple type *has the form $\tau = \langle \tau_1, ..., \tau_k \rangle$, where each τ_i is an element type, an ID type, a time type*[3] *(such as Date and Time), or an amount type. The members of τ are precisely those tuple objects $\langle x_1, ..., x_k \rangle$ where each x_i is a member of τ_i. For example, weblog sequences can be constructed from the tuple of $\langle Date, Time, URL \rangle$, where URL is a finite set of URLs.*

■

Clearly, using set types and tuple types one can define types for relations.

[2] In the literature the two terms of "sequence pattern" and "sequential pattern" have been used as synonyms. We will also use them interchangeably in this text. It should be noted that, except in Chapter 2 we use these terms in a more general sense.

[3] The domains of Date, Time, and Amount are defined in the natural way.

Definition 1.2. *A* sequence *over an element type τ is an ordered list*[4] $S = s_1...s_m$, *where*

- *each s_i (which can also be written as $S[i]$) is a member of τ, and is called an* element *of S;*
- *m is referred to as the* length *of S and is denoted by $|S|$;*
- *each number between 1 and $|S|$ is a* position *in S.*

A consecutive interval of sequence positions of the form $[i, j]$, where $1 \leqslant i \leqslant j \leqslant m$ is a window *of the sequence; $j - i + 1$ is referred to as the* length *of the window.* ■

Parenthesis and commas may be added to make sequences more readable.

Example 1.3. DNA sequences such as those shown in Figure 1.1 are sequences over $\{A, C, G, T\}$. The DNA sequence $S = ATGTATA$ has length 7, each number between 1 and 7 is a position in S, and $S[3]$ is the letter G.

Protein sequences such as those shown in Figure 1.2 are sequences over $\{A, C, D, E, F, G, H, I, K, L, M, N, P, Q, R, S, T, V, W, Y\}$.

Weblogs such as those shown in Figure 1.3 are over $\langle Date, Time, URL\rangle$.

Customer purchase histories such as those shown in Figure 1.4 are over $\tau = \langle CustomerID, Date, Time, Location, 2^I\rangle$, where the domain of *Location* is a (simple type) set of locations, and I is the set of product items. Storewide sales histories are similar. ■

The order among the elements of a sequence may be implied by time order as in event histories, or by physical positioning as in biological sequences.

The following general concepts are frequently used in biological sequence analysis:

- A *site* in a sequence (as in transcription binding site) is a short sequence window having some special biological property/interest. A site can be described by a start position and window length, or just a position. A site is usually characterized by the presence of some sequence pattern.
- Given a sequence $S = s_1...s_m$ and a position i of S, the prefix $s_1...s_{i-1}$ is often referred to as the *upstream* of i and the suffix $s_{i+1}...s_m$ is referred to as the *downstream* of i. The concepts are defined similarly for a window $[i, j]$ (or site) of S, with $s_1...s_{i-1}$ as the *upstream*, and $s_{j+1}...s_m$ as the *downstream*. It is common to refer to position $i - k$ of S as the $-k$ region of position i, and to refer to position $i + k$ as $+k$ region of position i.

1.2.2 Characteristics of Sequence Data

Sequence data have several distinct characteristics, which lead to many opportunities, as well as challenges, for sequence data mining. These include the following:

[4] Mathematically, an *ordered list* $s_1...s_m$ over an element type τ is defined to be a function from $\{1..m\}$ to τ, where m is some positive integer.

- Sequences can be very long (and hence sequence datasets can have very high dimensionality), and different sequences in a given application may have a large variation in lengths. For example, the length of a gene can be as large as over 100K, and as small as several hundreds.
- Absolute positions in sequences may/may not have significance. For example, sequences may need to be aligned based on their absolute positions and there can be a penalty on position changes through insertion/deletions. In certain situations, one may just want to look for patterns which can occur anywhere in the sequences.
- The relative ordering/positional relationship between elements in sequences is often important. In sequences, the fact that one element occurs to the left of another is usually different from the fact that the first element occurs to the right of the second. Moreover, the distance between two elements is also often significant. The relative ordering/positional relationship between elements is unique to sequences, and is not a factor for relational data or other high dimensional data such as microarray gene expression data.
- Patterns can be substrings or subsequences. Sometimes a pattern must occur as a substring (of consecutive elements) in a sequence, without gaps between elements. At other times, the elements in a pattern can occur as a subsequence (allowing gaps between matching elements) of a sequence.

1.2.3 Sequence Patterns and Sequence Models

We now discuss sequence patterns, sequence models[5], and related topics such as pattern matching and pattern support in sequence data. Due to the characteristics of sequence data discussed above, there are many possibilities for defining sequence patterns and sequence models. The purpose of this section is to provide a high-level unifying overview and show the many possibilities, rather than the detailed instances, of sequence patterns and sequence models. The detailed instances will be discussed in the subsequent chapters.

Roughly speaking, a sequence pattern/model consists of a number of single-position patterns plus some inter-positional constraints. A single-position pattern is essentially a condition on the underlying element type. A sequence pattern may contain zero, one, or multiple single-position patterns for each position, where the single-position patterns for a given position are perhaps associated with a probability distribution; inter-positional constraints specify certain linkage between positions; such linkage can include conditions on position distance, and perhaps also include transition probabilities from position to position when two or more single-position patterns are present for some position. Below we give more details on these variations, together with some examples.

---

[5] We choose to use the word pattern to mean a condition on a subset of the underlying data, and use the word model to mean a condition on all of the underlying data.

- A single-position pattern is a condition on the underlying element type defined recursively as follows: If τ is an item type, then a condition on τ can be "?" or "*" or "." (all denoting a single position wildcard or don't care), an element of τ, a subset of τ, or an interval of τ when τ is an ordered type. If τ is a set type of the form $\{\psi\}$, then a condition on τ is a finite set of conditions on ψ. If τ is a tuple type of the form $\langle\tau_1, ..., \tau_k\rangle$, then a condition on τ is an expression of the form $\langle c_1, ..., c_k\rangle$, where c_i is a condition on τ_i. Since patterns are used to capture behavior in data, it may not make sense to have non-? conditions on ID types.
For example, if $\tau = \langle\{A, B, C, D\}, \{E, F, G\}, int, real\rangle$, then a single-position condition can be $\langle A, \{E, G\}, ?, (20, 45]\rangle$. If $\tau = \{A, C, G, T\}$, then a single-position condition can be ?, C, $\{A, C\}$ etc.
While it is possible to use the Boolean operators "AND" and "OR" to construct more complex conditions, this is seldom done since data mining of patterns must deal with a huge search space even without these Boolean operators. The intervals for ordered attributes are usually determined through a binning/discretization process.
- A sequence pattern is a finite set of single-position patterns of the form $\{c_1, ..., c_k\}$, together with a description of the positional distance relationships on the c_i's and some other optional specifications. This formalization is general enough to include frequent sequence patterns, periodic patterns, sequence profile patterns, and Markov models. Below we give an overview of each of these.
A first representative sequence pattern type is the frequent sequence patterns. Each such a pattern consists of one single-position pattern for each position. For DNA sequences, an example of such a pattern is ATC. In the simplest case, the positions of the single-position patterns are a consecutive range of the positive integers – this is assumed when nothing is said about the relationships between the positions; in general, constraints on the positions (often referred to as gap constraints) can be included. For example, for the simplest case, A, T and C are at consecutive positions so that T's position is after A's position and C's position is after T's; for the general case, we may say that T's position is at least 2 and at most 5 positions after the position of A, and that C's position is at most 3 positions after T's position. One can also add a window constraint to restrict the difference between the positions of the last and the first single-position patterns to be at most, for example, 7.
Frequent sequence patterns can be viewed as periodic sequence patterns. We will discuss some distinctions between frequent sequence patterns and periodic sequence patterns below.
A second representative sequence pattern type is the sequence profile patterns. Such a pattern is over a set of positions, and it consists of a set of single-position pattern plus a probability distribution. Examples will be given in Chapter 4.

A third representative sequence pattern type is the Markov models. Such a model consists of a number of states plus probabilistic transitions between states. In some cases each state is also associated with a symbol emission probability distribution. Examples will be given in Chapter 4.
A fourth representative sequence pattern type is the partial order models. Each such a model contains a set of single-position patterns associated with a partial order on these patterns. In a sense, the position distance between pairs of single-position patterns is in the range of $[1, \infty)$. Such a model can capture a temporal event ordering on the events. Examples will be given in Chapter 5.

- In addition to sequence pattern mining discussed above, classification and clustering are also useful data mining tasks for sequence data. Neither these tasks nor their products fall under the general definition of sequence patterns given above. The characteristics of sequence data lead to new questions for these two tasks. For example, there are more possibilities for feature construction from sequence data. Moreover, in sequence data one may want to predict the "class" of a location in a long sequence, which does not have a counterpart for conventional relational/vector data. More details will be provided in Chapters 3 and 4.

We now turn to the issues regarding pattern matching and sequence pattern support in sequence data. We first need several definitions.

A *match* between a sequence pattern $p = p_1...p_k$ and a sequence $s = s_1...s_n$ is a function f from $\{1, ..., k\}$ to $\{1, ...m\}$ such that the condition p_i is satisfied in $s_{f(i)}$ and the associated constraints on p are satisfied. The concept of satisfaction is defined in the natural manner.

For each match between a sequence pattern and a sequence, let the *match interval* be defined as $[low, high]$, where low is the smallest position, and $high$ is the largest position, in the sequence for the match. We note that, for sequence patterns with gaps, it is possible that the matching interval of one match is properly contained in the matching interval of a second match.

Several possibilities exist regarding which matches can contribute towards the count/support of a pattern:

- One sequence contributes at most one match and the support/count of pattern is with respect to the whole dataset. This simple case is very similar to the conventional transactional data case.
- One sequence contributes multiple matches and the count of pattern is with respect to one sequence. Three options exist: (b) Different contributing matches are completely disjoint, in the sense that the matching intervals of different contributing matches must be completely disjoint. (b) Different contributing matches are sufficiently disjoint, in the sense that the matching intervals of different contributing matches must not overlap more than some given threshold. (c) All matches are counted. For options (a) and (b), it may be computationally expensive to determine the highest possible number of matches of a pattern in a sequence.

A sequence model can be used as a generative device. For example, one can compute the most likely sequence that can be generated by a Markov model.

Some distinctions can be made between sequence patterns and sequence models, similar to the distinctions between general patterns and general models. A pattern is usually *partial* (or local) in the sense that it may occur only in a subset of the sequences under consideration. On the other hand, a model is usually *total* (or global) in the sense that it can be applied to every sequence under consideration.

1.3 General Data Mining Processes and Research Issues

In this section we give a brief high level overview of the general data mining process, and the general issues of interest in data mining research and applications. More details on these can be found in general data mining texts.

The typical steps of the data mining process are the following:

- Understanding the application requirements and the data. In this step the analyst will need to understand what is important, and how such importance is reflected in data.
- Preprocessing of the data by data cleaning, feature/data selection, and data transformation. Data cleaning is concerned with removing inconsistency in data, with integrating data from heterogeneous sources etc. Feature selection is concerned with selecting the more useful features (for a particular data mining task) from a large number of candidate features. Feature construction is about producing new features from existing features. Data transformation is concerned with mapping data from one form to another. Discretization (also called binning) is a common approach of data transformation, where one maps an attribute with a large domain into an attribute with a smaller domain. Common discretization methods include equi-width binning, equi-density binning, and entropy-based binning.
- Mining the patterns/models. This is done by running some data mining algorithms on the data produced from the last step above.
- Evaluation of the mining result. In this step the data analyst will apply various measures to evaluate the goodness of the mined patterns or models for the application under consideration.

These steps may be iterated to improve the quality of the mining result. Improvement is possible since one's understanding of the data/application deepens after one or more iterations of working through the data.

Naturally, data mining research should address issues of practical/theoretical interest, and solving important problems, in data mining applications. Data mining research often considers the following technical issues:

- Formulating useful new concepts that have high potential to lead to advances of research in the field.
- Designing novel techniques for efficiency and scalability in computational space/time, for dealing with large volume of data and with high dimensionality of data. The techniques should address the unique challenges and take advantage of the unique opportunities of the underlying application/data.
- Optimizing cluster/classification quality under measures such as accuracy, precision and recall, and cluster quality (intra-cluster similarity and inter-cluster dissimilarity).
- Optimizing pattern interestingness under appropriate measures, such as support/confidence, surprise, lift/novelty and actionability.

Details on the concepts discussed above, together with examples on the design of techniques and on various optimizations, will be given in later chapters.

1.4 Overview of the Book

The rest of this book is organized as follows:

Chapter 2 first motivates and defines the task of sequential pattern mining. Then, it discusses two essential kinds of methods: the Apriori-like, breadth-first search methods and the pattern-growth, depth-first search methods. It also discusses constrained sequential pattern mining techniques, and closed sequential pattern mining. Constrained mining allows a user to get a specific subset of sequential patterns instead of all patterns by specifying certain constraints. Closed sequential patterns are useful for removing certain redundancy in the set of sequential patterns and hence for producing smaller sets of mined patterns without loss of information.

Chapter 3 is concerned with the classification and clustering of sequence data. It first provides a general categorization of sequence classification and sequence clustering. There are three general tasks. Two of those tasks are concerned with whole sequences and will be presented there. The third topic, namely sequence motifs (site/position-based identification and characterization of sequence families), is presented in Chapter 4. Chapter 3 also contains two sections on sequence features (concerning various feature types and general feature selection criteria) and sequence similarity/distance functions. These materials will be useful for not only classification and clustering, but also other topics (such as identification and characterization of sequence families).

Chapter 4 is concerned with sequence motifs. It includes the discussion on motif finding and the use of motifs in sequence analysis. A motif is essentially a short distinctive sequence pattern shared by a number of related sequences. The motif finding task is concerned with site-focused identification and characterization of sequence families. It can be viewed as a hybrid of clustering and classification, and is an iterative process. Motif analysis is concerned with predicting whether sequences match a certain motif, and the sequence position where a match occurs. This chapter will present some major motif representations. While there have been many algorithms for motif finding and motif analysis, most of them are instances of one of the following three algorithmic approaches, namely dynamic programming, expectation maximization, and Gibbs sampling. This chapter also presents these algorithmic approaches.

Chapter 5 considers the mining of partial orders from sequence data. Partial orders can be used to capture the preferred ordering among events. We introduce two types of mining tasks and their methods. First, we discuss a method to mine frequent closed partial orders from strings, which can be regarded as the generalization of sequential pattern mining. Second, we discuss how to find the best partial order that is shared by the majority in a set of sequences, which can be modeled as an optimization problem.

Chapter 6 considers the mining of distinguishing sequence patterns and rare events from sequence data. A distinguishing sequence pattern is a sequence pattern that (i) characterizes a family of sequences and distinguishes the family from other sequences, or (ii) characterizes a special site of sequences and distinguishes the site from other parts of sequences, or (iii) signals something unusual about some sequences. This chapter first discusses four types of distinguishing sequence patterns, and then gives some methods/algorithms for the mining of two of those types. (The other two types were discussed in Chapter 4.) Distinguishing sequence patterns are also useful as candidate sequence features.

Chapter 7 provides a brief overview of related research topics, including partial periodic pattern mining, structured-data mining (containing sequence mining, tree mining, graph mining and time series mining as special cases) and bioinformatics. It also briefly discusses sequence alignment, which is needed for understanding certain materials in several other chapters. Finally, it provides some pointers to biological sequence databases and resources.

We try to make each chapter essentially self-contained. That is, each chapter can be read independently. Terms are indexed in the appendix to facilitate cross referencing.

2

Frequent and Closed Sequence Patterns

Sequential pattern mining [3] is an essential task in sequence data mining. In this chapter, we first motivate the task of sequential pattern mining. Then, we discuss two kinds of major methods: the Apriori-like, breadth-first search methods and the pattern-growth, depth-first search methods. More often than not, a user may want some specific subset of sequential patterns instead of all patterns. This application requirement can be addressed by the constrained sequential pattern mining techniques. When mining a large database, there can be many sequential patterns. Redundancy may exist among sequential patterns. We discuss mining closed sequential patterns which can remove some redundancy.

It should be noted that the "sequence patterns" considered in this chapter are a special class of patterns in sequence data. For historical reasons and the lack of a better name, we will still call them "sequence patterns."

2.1 Sequential Patterns

Example 2.1 (Sequential patterns and applications). eShop sells technology products. Each customer shopping at eShop has a customer-id. A customer transaction contains a set of products bought by the customer at some time point. eShop maintains a sequence database which contains all transactions that happened at eShop.

An important marketing approach of eShop is to send promotion advertisements to targeted customers. In order to design attractive promotions to be sent to relevant customers, it is a good idea to utilize patterns in historical data. Transaction information can be used to construct the shopping history sequences of customers: For each customer, we collect all transactions of the customer and form a sequence in the transaction time-stamp ascending order. For example, some sequences are shown in Table 2.1.

A sequence consists of a number of transactions. For example, sequence $C1$ of Table 2.1 has 5 transactions. In the first transaction, the customer buys only

Customer-id	Transaction sequence
$C1$	$a(abc)(ac)d(cf)$
$C2$	$(ad)c(bc)(ae)$
$C3$	$(ef)(ab)(df)cb$
$C4$	$eg(af)cbc$

Table 2.1. A customer transaction sequence database.

one product, a. In the second transaction, the customer buys three products, namely a, b and c. A product may appear more than once in a sequence. For example, in sequence $C1$, product a appears in the first three transactions.

Can we find some patterns in the sequence database that can help us to capture the common purchase patterns? Frequent subsequences as sequential patterns are particularly useful. For example, $(ab)dc$ is a subsequence of $C1$ and $C3$. If the support threshold is set to 2, that is, a sequential pattern should be a subsequence of at least 2 sequences in the database, then $(ab)dc$ is a sequential pattern.

Sequential patterns can be used in two aspects in this application. First, sequential patterns capture the common purchase patterns of customers. For example, sequential pattern $(ab)dc$ tells that at least 2 customers buy products a and b in a transaction, then buy d in a later transaction, and then buy c (after buying d). In the context of marketing campaign design, sequential patterns can be used to design promotions. For example, suppose c is a highly profitable product and d is an inexpensive one. Then, knowing sequential pattern $(ab)dc$, one may promote product d to attract customers to buy c in sequence.

Second, sequential patterns can also be used for predicting behavior of individual customers. For example, if $(ab)dc$ is a sequential pattern, we can send advertisement and promotions of c to all customers who bought a, b and d before in sequence, since they may buy c in a future transaction. ■

Sequential patterns are useful in many other applications in addition to marketing, such as web log mining and web page recommendation systems, bio-sequence analysis, medical treatment sequence analysis, and safety management and disaster prevention.

Now, let us define the problem of sequential pattern mining formally.

Let $I = \{i_1, i_2, \ldots, i_n\}$ be a set of *items*. An *itemset* is a subset of items. A *sequence* is an ordered list of itemsets. A sequence s is denoted by $s_1 s_2 \cdots s_l$, where each s_j $(1 \leqslant j \leqslant l)$ is an itemset. s_j is also called an *element* or a *transaction* of the sequence; a transaction is denoted as $(x_1 x_2 \cdots x_m)$, where x_k $(1 \leqslant k \leqslant m)$ is an item. For the sake of brevity, the brackets are omitted if an element has only one item, that is, element (x) is written as x. An item can occur at most once in an element of a sequence, but can occur multiple times in different elements of a sequence. The number of instances of items in a sequence is called the *i-length* of the sequence[1]. A sequence with i-length l is called an *l-sequence.* A sequence $\alpha = a_1 a_2 \cdots a_n$ is called a *subsequence* of another sequence $\beta = b_1 b_2 \cdots b_m$ and β a *super-sequence* of α, denoted as $\alpha \sqsubseteq \beta$, if there exist integers $1 \leqslant j_1 < j_2 < \cdots < j_n \leqslant m$ such that $a_1 \subseteq b_{j_1}$, $a_2 \subseteq b_{j_2}$, ..., $a_n \subseteq b_{j_n}$.

A *sequence database* S is a set of tuples of the form (sid, s), where sid is a *sequence_id* and s a sequence. A tuple (sid, s) is said to *contain* a sequence α, if α is a subsequence of s. The *support* of a sequence α in a sequence database S is the number of tuples in the database containing α, that is,

$$support_S(\alpha) = \mid \{(sid, s) | (sid, s) \in S) \wedge (\alpha \sqsubseteq s)\} \mid$$

It can be denoted as *support*(α) if the sequence database is clear from the context.

Given a positive integer *min_support* as the *support threshold*, a sequence α is called a *sequential pattern* in sequence database S if $support_S(\alpha) \geqslant$ *min_support*. A sequential pattern with i-length l is called an *l-pattern.*

Example 2.2 (Running example). Let our running sequence database be S given in Table 2.1 and *min_support* $= 2$. The set of items in the database is $\{a, b, c, d, e, f, g\}$.

A sequence $a(abc)(ac)d(cf)$ has five elements: (a), (abc), (ac), (d) and (cf), where items a and c appear more than once, respectively, in different elements. It is a 9-sequence since there are 9 instances appearing in that sequence. Item a occurs three times in this sequence, so it contributes 3 to the i-length of the sequence. However, the whole sequence $a(abc)(ac)d(cf)$ contributes only one to the support of a. Also, sequence $a(bc)df$ is a subsequence of $a(abc)(ac)d(cf)$. Since sequences $C1$ and $C3$ are the only two sequences containing subsequence $s = (ab)c$, s has a support of 2; so s is a sequential pattern of i-length 3 (that is, a 3-pattern). ■

Given a sequence database and a minimum support threshold, the *sequential pattern mining* problem is to find the complete set of sequential patterns in the database.

The sequential pattern mining problem [3] was also simultaneously identified as the frequent episode mining problem by Mannila et al. [75]. Frequent episode mining can be generalized to frequent partial order mining which will be discussed in Chapter 5.

---

[1] Observe that the i-length of a sequence is the total number of items in the sequence, whereas the length of the sequence is the total number of positions in the sequence. The two concepts are equivalent if each sequence position has just a single item.

2.2 GSP: An Apriori-like Method

Sequential patterns have a monotonic property. For example, $(ab)dc$ is a sequential pattern in Example 2.2. Then, all subsequences of the pattern, namely a, b, d, c, (ab), ad, ac, bd, bc, dc, $(ab)d$, $(ab)c$, adc, and bdc, are sequential patterns as well. The reason is that every sequence in the database containing $(ab)dc$ also (trivially) contains every subsequence.

Theorem 2.3 (Apriori property). *For a sequence s and a subsquence s' of s, $support(s) \leqslant support(s')$. Moreover, if s is a sequential pattern, so is s'.*

Proof. Consider each sequence seq in the sequence database in question such that s is a subsequence of seq. Clearly, s' must be also a subsequence of seq since s' is a subsequence of s. Therefore, the number of sequences that contain s' cannot be less than the number of sequences that contain s. That is, $support(s) \leqslant support(s')$.

If s is a sequential pattern, then $support(s) \geqslant min\_support$, where $min\_support$ is the minimum support threshold. Therefore, $support(s') \geqslant support(s) \geqslant min\_support$. That is, s' is also a sequential pattern. ∎

Using the *Apriori* property, we can develop breadth-first search algorithms to find all sequential patterns. The general idea is that, if s is not a sequential pattern, we do not search any super-sequence of s.

A typical sequential pattern mining method, *GSP* [101], mines sequential patterns by adopting a candidate subsequence generation-and-test approach based on the *Apriori* property. The method is illustrated in the following example.

Example 2.4 (GSP). Given the database S and the minimum support threshold $min\_support$ in Example 2.2, *GSP* first scans S, collects the support for each item, and finds the set of frequent items, that is, frequent length-1 subsequences (in the form of "*item:support*"): $a:4, b:4, c:3, d:3, e:3, f:3, g:1$.

By filtering out the infrequent item g, we obtain the first seed set

$$L_1 = \{a, b, c, d, e, f\},$$

where each member of L_1 represents a 1-element sequential pattern. Each subsequent pass starts with the seed set found in the previous pass and uses it to generate new potential sequential patterns, called *candidate sequences.*

From L_1 (a set containing 6 length-1 sequential patterns), we generate the following set of $6 \times 6 + \frac{6\times5}{2} = 51$ candidate sequences:

$$C_2 = \{aa, ab, \ldots, af, ba, bb, \ldots, ff, (ab), (ac), \ldots, (ef)\}.$$

Then, the sequence database is scanned again, and the supports of sequences in C_2 are counted. Those sequences in C_2 passing the minimum

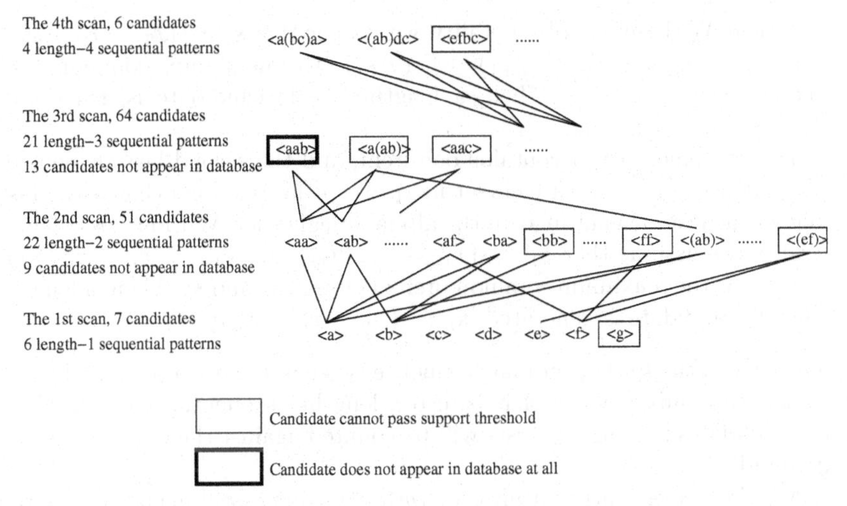

Fig. 2.1. Candidates and sequential patterns in *GSP*.

support threshold are the length-2 sequential patterns. Using the length-2 sequential patterns, we can generate C_3, the set of length-3 candidates.

The multi-scan mining process is shown in Figure 2.1. The set of candidates is generated by a self-join of the sequential patterns found in the previous pass. In the k-th pass, a sequence is a candidate only if each of its length-$(k-1)$ subsequences is a sequential pattern found at the $(k-1)$-th pass. A new scan of the database collects the support for each candidate sequence and finds the new set of sequential patterns. This set becomes the seed for the next pass. The algorithm terminates when no sequential pattern is found in a pass, or when no candidate sequence is generated. Clearly, the number of scans is at least the maximum i-length of sequential patterns. It needs one more scan if the sequential patterns obtained in the last scan lead to the generation of new candidates. ■

While the general procedure of *GSP* is easy to understand, the candidate-generation in the algorithm is non-trivial. Generally, we can list all items in a transaction of a sequence in the alphabetical order. Suppose that s_1 and s_2 are two length-k sequential patterns ($k \geqslant 1$) such that s_1 and s_2 are identical except for the last element. Then, s_1 and s_2 are used to generate a length-$(k+1)$ candidate if one of the following situations happens.

- The last element of s_1 contains only one item, and so does the last element of s_2. Without loss of generality, we assume that $s_1 = sx$ and $s_2 = sy$ where s is the maximum common prefix of s_1 and s_2, and x, y are two items. Then, the following three length-$(k+1)$ candidates are generated: $s(xy)$, sxy and syx.
- The last elements of s_1 and s_2 contain more than one item, and except for the last item in the alphabetical order, the last itemsets of s_1 and s_2 are

identical. Without loss of generality, we assume that $s_1 = s(x_1 \cdots x_{m-1}x_m)$ and $s_2 = s(x_1 \cdots x_{m-1}x_{m+1})$, where s is the maximum common prefix between s_1 and s_2. Then, a length-$(k+1)$ candidate is generated: $s(x_1 \cdots x_{m-1}x_m x_{m+1})$.

- The last element of s_2 contains one item, and the second last element of s_2 is identical to the last element in s_1 except for one item that is the last one in the last element in s_1 in the alphabetical order. Without loss of generality, we assume that $s_1 = s(x_1 \cdots x_{m-1}x_m)$ and $s_2 = s(x_1 \cdots x_{m-1})y$, where s is the maximum common prefix between s_1 and s_2. Then, a length-$(k+1)$ candidate is generated: $s_1 = s(x_1 \cdots x_{m-1}x_m)y$.

Once a length-$(k+1)$ candidate sequence is generated, we also test whether every length-k subsequence of it is also a length-k sequential pattern. Only those candidates passing the tests will be counted against the database in the next round.

GSP, though benefiting from the *Apriori* pruning, still generates a large number of candidates. In Example 2.4, 6 length-1 sequential patterns generate 51 length-2 candidates, 22 length-2 sequential patterns generate 64 length-3 candidates, and so on. Some candidates generated by *GSP* may not appear in the database at all. For example, 13 out of 64 length-3 candidates do not appear in the database.

In addition to *GSP*, some other *Apriori*-like, breadth-first search methods have been developed. For example, *PSP* [77] improves *GSP* by exploiting an intermediate data structure. *SPADE* [132] uses a vertical id-list format, and also divides a sequence lattice into small parts.

2.3 PrefixSpan: A Pattern-growth, Depth-first Search Method

In addition to the *Apriori*-like, breadth-first search methods, pattern-growth, depth-first search methods are a category of more efficient approaches for sequential pattern mining. We first analyze the overhead of *Apriori*-like, breadth-first search methods. Then, we introduce *PrefixSpan*, a representative of the pattern-growth, depth-first search methods.

2.3.1 Apriori-like, Breadth-first Search versus Pattern-growth, Depth-first Search

The *Apriori*-like, breadth-first search methods bear three kinds of nontrivial, inherent cost which are independent of detailed implementation techniques.

- **Potentially huge sets of candidate sequences**. Since the set of candidate sequences includes all the possible permutations of the elements and repetition of items in a sequence, an *Apriori*-like, breadth-first search method may generate a large set of candidate sequences even for a moderate seed set. For example, if there are $1,000$ length-1 sequential patterns $a_1, a_2, \ldots, a_{1000}$, an *Apriori*-like algorithm will generate

$$1000 \times 1000 + \frac{1000 \times 999}{2} = 1,499,500$$

length-2 candidate sequences. (The first term is derived from the set of candidate sequences $\{a_1a_1, a_1a_2, \ldots, a_1a_{1000}, a_2a_1, a_2a_2, \ldots, a_{1000}a_{1000}\}$, and the second term is derived from the set of candidate sequences $\{(a_1a_2), (a_1a_3), \ldots, (a_{999}a_{1000})\}$.)

- **Multiple scans of databases.** Since each database scan considers sequences whose i-length is one larger than that of the previous scan, to find a sequential pattern $(abc)(abc)\ (abc)(abc)(abc)$, an *Apriori*-like method must scan the database at least 15 times.
- **Difficulties at mining long sequential patterns.** A long sequential pattern must grow from a combination of short ones, but the number of such candidate sequences is exponential to the i-length of the sequential patterns to be mined. For example, suppose there is only a single sequence of length 100, $a_1a_2 \ldots a_{100}$, in the database, and the min_support threshold is 1 (that is, every occurring pattern is frequent), to (re-)derive this length-100 sequential pattern, an *Apriori*-like method has to generate 100 length-1 candidate sequences, $100 \times 100 + \frac{100 \times 99}{2} = 14,950$ length-2 candidate sequences, $\binom{100}{3} = 161,700$ length-3 candidate sequences[2], ... Obviously, the total number of candidate sequences to be generated is greater than $\sum_{i=1}^{100} \binom{100}{i} = 2^{100} - 1 \approx 10^{30}$.

In many applications, it is not unusual to encounter a large number of sequential patterns and long sequences.

In this section, we introduce *PrefixSpan* [88], a pattern-growth, depth-first search method. The general ideas of *PrefixSpan* are the following three points.

- Instead of generating a large number of candidates, *PrefixSpan* preserves in some compressed forms the essential groupings of the original data elements for mining. Then the analysis is focused on counting the frequency of the relevant data sets instead of the candidate sets.
- Instead of scanning the entire database to match against the whole set of candidates in each pass, *PrefixSpan* partitions the data set to be examined as well as the set of patterns to be examined by database projection. Such a divide-and-conquer methodology substantially reduces the search space and leads to high performance.

[2] Notice that *Apriori* does cut a substantial amount of search space. Otherwise, the number of length-3 candidate sequences would have been $100 \times 100 \times 100 + 100 \times 100 \times 99 + \frac{100 \times 99 \times 98}{3 \times 2} = 2,151,700$.

- With the growing capacity of main memory and the substantial reduction of database size by database projection as well as the space needed for manipulating large sets of candidates, a substantial portion of data can be put into main memory for mining. Pseudo-projection has been developed for pointer-based traversal. Reported performance studies have shown the effectiveness of such techniques.

2.3.2 PrefixSpan

Let us first introduce the concepts of prefix and suffix which are essential in *PrefixSpan*.

Definition 2.5 (Prefix). *Suppose all the items within an element are listed alphabetically. For a given sequence* $\alpha = e_1 e_2 \cdots e_n$*, where each* e_i $(1 \leqslant i \leqslant n)$ *is an element, a sequence* $\beta = e_1' e_2' \cdots e_m'$ $(m \leqslant n)$ *is called a* **prefix** *of* α *if (1)* $e_i' = e_i$ *for* $i \leqslant m - 1$*; (2)* $e_m' \subseteq e_m$*; and (3) all items in* $(e_m - e_m')$ *are alphabetically after those in* e_m'. ■

For example, consider sequence $s = a(abc)(ac)d(cf)$. Sequences a, aa, $a(ab)$ and $a(abc)$ are prefixes of s, but neither ab nor $a(bc)$ is a prefix.

Definition 2.6 (Suffix). *Consider a sequence* $\alpha = e_1 e_2 \cdots e_n$*, where each* e_i $(1 \leqslant i \leqslant n)$ *is an element. Let* $\beta = e_1' e_2' \cdots e_{m-1}' e_m'$ $(m \leqslant n)$ *be a subsequence of* α*. Sequence* $\gamma = e_l'' e_{l+1} \cdots e_n$ *is the* **suffix** *of* α *with respect to prefix* β*, denoted as* $\gamma = \alpha / \beta$*, if*

1. $l = i_m$ *such that there exist* $1 \leqslant i_1 < \cdots < i_m \leqslant n$ *such that* $e_j' \subseteq e_{i_j}$ $(1 \leqslant j \leqslant m)$*, and* i_m *is minimized. In other words,* $e_1 ... e_l$ *is the shortest prefix of* α *which contains* $e_1' e_2' \cdots e_{m-1}' e_m'$ *as a subsequence; and*
2. e_l'' *is the set of items in* $e_l - e_m'$ *that are alphabetically after all items in* e_m'.

If e_l'' *is not empty, the suffix is also denoted as* $(\_$ *items in* $e_l'')e_{l+1} \cdots e_n$*. Note that if* β *is not a subsequence of* α*, the suffix of* α *with respect to* β *is empty.* ■

Example 2.7 (Prefix and suffix). In our running example, for the sequence $s = a(abc)(ac)d(cf)$, $(abc)(ac)d(cf)$ is the *suffix* with respect to a, $(\_c)(ac)d(cf)$ is the *suffix* with respect to ab, and $(ac)d(cf)$ is the *suffix* with respect to (ac). ■

Based on the concepts of prefix and suffix, the problem of mining sequential patterns can be decomposed into a set of subproblems as follows.

1. Let $\{x_1, x_2, \ldots, x_n\}$ be the complete set of length-1 sequential patterns in a sequence database S. The complete set of sequential patterns in S can be divided into n disjoint subsets. The ith subset $(1 \leqslant i \leqslant n)$ is the set of sequential patterns with prefix x_i.

2. Let α be a length-l sequential pattern and $\{\beta_1, \beta_2, \ldots, \beta_m\}$ be the set of all length-$(l+1)$ sequential patterns with prefix α. The complete set of sequential patterns with prefix α, except for α itself, can be divided into m disjoint subsets. The jth subset $(1 \leqslant j \leqslant m)$ is the set of sequential patterns prefixed with β_j.

The above recursive partitioning of the sequential pattern mining problem forms a *divide-and-conquer* framework. The above partitioning process can be visualized as a *sequence enumeration tree* .

Example 2.8 (Sequence enumeration tree). Let the set of items $I = \{a, b, c, d\}$. Figure 2.2 shows a sequence enumeration tree which enumerates all possible sequences formed using the items.

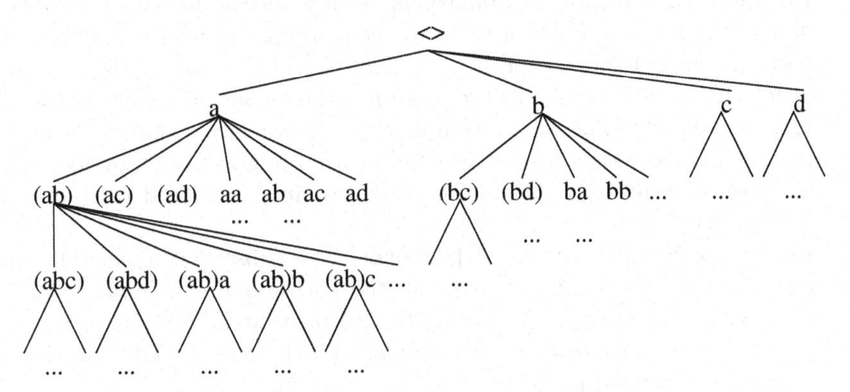

Fig. 2.2. The sequence enumeration tree on the set of items $\{a, b, c, d\}$.

The divide-and-conquer partitioning process in *PrefixSpan* is to conduct a depth-first search of the sequence enumeration tree. ■

To mine the subsets of sequential patterns, the corresponding projected databases can be constructed.

Definition 2.9 (Projected database). *Let α be a sequential pattern in a sequence database S. The α-**projected database**, denoted as $S|_\alpha$, is the collection of suffixes of sequences in S with respect to prefix α.* ■

Let us examine how to use the prefix-based projection approach to mine sequential patterns.

Example 2.10 (PrefixSpan). For the same sequence database S in Table 2.1 with $min\_sup = 2$, sequential patterns in S can be mined by a prefix-projection method in the following steps.

prefix	projected (suffix) database	sequential patterns
a	$(abc)(ac)d(cf)$, $(\_d)c(bc)(ae)$, $(\_b)(df)cb$, $(\_f)cbc$	a, aa, ab, $a(bc)$, $a(bc)a$, aba, abc, (ab), $(ab)c$, $(ab)d$, $(ab)f$, $(ab)dc$, ac, aca, acb, acc, ad, adc, af
b	$(\_c)(ac)d(cf)$, $(\_c)(ae)$, $(df)cb$, c	b, ba, bc, (bc), $(bc)a$, bd, bdc, bf
c	$(ac)d(cf)$, $(bc)(ae)$, b, bc	c, ca, cb, cc
d	(cf), $c(bc)(ae)$, $(\_f)cb$	d, db, dc, dcb
e	$(\_f)(ab)(df)cb$, $(af)cbc$	e, ea, eab, eac, $eacb$, eb, ebc, ec, ecb, ef, efb, efc, $efcb$.
f	$(ab)(df)cb$, cbc	f, fb, fbc, fc, fcb

Table 2.2. Projected databases and sequential patterns

1. **Find length-1 sequential patterns.** Scan S once to find all the frequent items in sequences. Each of these frequent items is a length-1 sequential pattern. They are $a:4$, $b:4$, $c:4$, $d:3$, $e:3$, and $f:3$, where the notation "*pattern* : *count*" represents the pattern and its associated support count.
2. **Divide search space.** The complete set of sequential patterns can be partitioned into the following six subsets according to the six prefixes: (1) the ones with prefix a, (2) the ones with prefix b, ..., and (6) the ones with prefix f.
3. **Find subsets of sequential patterns.** The subsets of sequential patterns can be mined by constructing the corresponding set of *projected databases* and mining each recursively. The projected databases as well as sequential patterns found in them are listed in Table 2.2, while the mining process is explained as follows.
   a) **Find sequential patterns with prefix** a. Only the sequences containing a should be collected. Moreover, in a sequence containing a, only the subsequence prefixed with the first occurrence of a should be considered. For example, in sequence $(ef)(ab)(df)cb$, only the subsequence $(\_b)(df)cb$ should be considered for mining sequential patterns prefixed with a. Notice that $(\_b)$ means that the last element in the prefix, which is a, together with b, form one element.
   The sequences in S containing a are projected with respect to a to form the *a-projected database*, which consists of four suffix sequences: $(abc)(ac)d(cf)$, $(\_d)c(bc)(ae)$, $(\_b)(df)cb$ and $(\_f)cbc$.
   By scanning the a-projected database once, its locally frequent items are $a:2$, $b:4$, $\_b:2$, $c:4$, $d:2$, and $f:2$. Thus all the length-2 sequential patterns prefixed with a are found, and they are: $aa:2$, $ab:4$, $(ab):2$, $ac:4$, $ad:2$, and $af:2$.
   Recursively, all sequential patterns with prefix a can be partitioned into 6 subsets: (1) those prefixed with aa, (2) those with ab, ..., and finally, (6) those with af. These subsets can be mined by constructing respective projected databases and mining each recursively as follows.

i. The aa-projected database consists of two non-empty (suffix) subsequences prefixed with aa: $\{(\_bc)(ac)d(cf)$, $\{(\_e)\}$. Since there is no hope to generate any frequent subsequence from this projected database, the processing of the aa-projected database terminates.
ii. The ab-projected database consists of the following three suffix sequences: $(\_c)(ac)d(cf)$, $(\_c)a$, and c. Recursively mining the ab-projected database returns four sequential patterns: $(\_c)$, $(\_c)a$, a, and c (that is, $a(bc)$, $a(bc)a$, aba, and abc.) They form the complete set of sequential patterns prefixed with ab.
iii. The (ab)-projected database contains the following two sequences: $(\_c)(ac)d(cf)$ and $(df)cb$, which leads to the finding of the following sequential patterns prefixed with (ab): c, d, f, and dc.
iv. The ac-, ad- and af- projected databases can be constructed and recursively mined similarly. The sequential patterns found are shown in Table 2.2.

b) **Find sequential patterns with prefix b, c, d, e and f, respectively**. This can be done by constructing the b-, c- d-, e- and f-projected databases and mining them respectively. The projected databases as well as the sequential patterns found are shown in Table 2.2.

4. **The set of sequential patterns is the collection of patterns found in the above recursive mining process**. One can verify that it returns exactly the same set of sequential patterns as what *GSP* does. ■

Based on the above discussion, the algorithm of *PrefixSpan* is presented in Figure 2.3.

Input: A sequence database S, and the minimum support threshold *min_support*.
Output: The complete set of sequential patterns.
Method: Call *PrefixSpan*$(\emptyset, 0, S)$.

Subroutine *PrefixSpan*$(\alpha, l, S|_\alpha)$
The parameters are (1) α is a sequential pattern; (2) l is the i-length of α; and (3) $S|_\alpha$ is the α-projected database if $\alpha \neq \emptyset$, otherwise, it is the sequence database S.
Method:

1. Scan $S|_\alpha$ once, find each frequent item b such that
   a) b can be assembled to the last element of α to form a sequential pattern; or
   b) b can be appended to α to form a sequential pattern.
2. For each frequent item b, append it to α to form a sequential pattern α', and output α';
3. For each α', construct α'-projected database $S|_{\alpha'}$, and call *PrefixSpan*$(\alpha', l+1, S|_{\alpha'})$.

Fig. 2.3. Algorithm *PrefixSpan*.

Let us analyze the efficiency of the algorithm.

- **No candidate sequence needs to be generated by *PrefixSpan*.** Unlike *Apriori*-like algorithms, *PrefixSpan* only grows longer sequential patterns from the shorter frequent ones. It neither generates nor tests any candidate sequence non-existent in a projected database. Comparing with *GSP*, which generates and tests a substantial number of candidate sequences, *PrefixSpan* searches a much smaller space.
- **Projected databases keep shrinking.** It is easy to see that a projected database is smaller than the original one because only the suffix subsequences of a frequent prefix are projected into a projected database. In practice, the shrinking factors can be significant because (1) usually, only a small set of sequential patterns grow quite long in a sequence database, and thus the number of sequences in a projected database usually reduces substantially when prefix grows; and (2) projection only takes the suffix portion with respect to a prefix.
- **The major cost of *PrefixSpan* is the construction of projected databases.** In the worst case, *PrefixSpan* constructs a projected database for every sequential pattern. If there exist a good number of sequential patterns, the cost is non-trivial. Techniques for reducing the number of projected databases will be discussed in the next subsection.

Theoretically, the problem of mining the complete set of sequential patterns is #P-complete [33]. Therefore, it is impossible to have a polynomial time algorithm unless $P = NP$. Even if $P = NP$, it is still unclear whether a polynomial time algorithm exists.

Interestingly, we can show that the *PrefixSpan* algorithm is pseudo-polynomial. That is, the complexity of *PrefixSpan* is linear with respect to the number of sequential patterns, since each projection generates at least one sequential pattern, and the projection cost is upper bounded by the time of scanning the database once, and counting frequent items in the suffixes.

2.3.3 Pseudo-Projection

The above analysis shows that the major cost of *PrefixSpan* is database projection, that is, forming projected databases recursively. Usually, a large number of projected databases will be generated in sequential pattern mining. If the number and/or the size of projected databases can be reduced, the performance of sequential pattern mining can be further improved.

One technique which may reduce the number and size of projected databases is *pseudo-projection*. The idea is outlined as follows. Instead of performing physical projection, one can register the index (or identifier) of the corresponding sequence and the starting position of the projected suffix in the sequence. Then, a physical projection of a sequence is replaced by registering a sequence identifier and the projected position index point. *Pseudo-projection* reduces the cost of projection substantially when the projected database can fit in main memory.

This method is based on the following observation. For any sequence s, each projection can be represented by a corresponding projection position (an index point) instead of copying the whole suffix as a projected subsequence. Consider a sequence $a(abc)(ac)d(cf)$. Physical projections may lead to repeated copying of different suffixes of the sequence. An index position pointer may save physical projection of the suffix and thus save both space and time of generating numerous physical projected databases.

Example 2.11. (Pseudo-projection) For the same sequence database S in Table 2.1 with $min\_sup = 2$, the sequential patterns in S can be mined by pseudo-projection method as follows.

Suppose the sequence database S in Table 2.1 can be held in main memory. Instead of constructing the a-projected database, one can represent the projected suffix sequences using pointer (sequence_id) and offset(s). For example, the projection of sequence $s_1 = a(abc)d(ae)(cf)$ with regard to the a-projection consists two pieces of information: (1) a *pointer* to s_1 which could be the string_id s_1, and (2) the *offset(s)*, which should be a single integer, such as 2, if there is a single projection point; and a set of integers, such as $\{2, 3, 6\}$, if there are multiple projection points. Each offset indicates at which position the projection starts in the sequence.

Sequence_id	Sequence	a	b	c	d	f	aa	...
10	$a(abc)(ac)d(cf)$	2, 3, 6	4	5, 7	8	\$	3, 6	...
20	$(ad)c(bc)(ae)$	2	5	4, 6	3	$\emptyset$	7	...
30	$(ef)(ab)(df)cb$	4	5	8	6	3, 7	$\emptyset$	...
40	$eg(af)cbc$	4	6	6	$\emptyset$	5	$\emptyset$	...

Table 2.3. A sequence database and some of its pseudo-projected databases

The projected databases for prefixes a-, b-, c-, d-, f-, and aa- are shown in Table 2.3, where \$ indicates the prefix has an occurrence in the current sequence but its projected suffix is empty, whereas $\emptyset$ indicates that there is no occurrence of the prefix in the corresponding sequence. From Table 2.3, one can see that the pseudo-projected database usually takes much less space than its corresponding physically projected one. ■

Pseudo-projection avoids physically copying suffixes. Thus, it is efficient in terms of both running time and space. However, it may not be efficient if the pseudo-projection is used for disk-based accessing since random access disk space is costly. Based on this observation, the suggested approach is the following: if the original sequence database or the projected databases are too big to fit into main memory, then physical projection should be applied;

however, the execution should be swapped to pseudo-projection once the projected databases can fit in main memory. This methodology is adopted in the *PrefixSpan* implementation.

Based on *PrefixSpan*, some more efficient pattern-growth, depth-first search methods have been developed recently. For example, Chiu et al. [19] propose a new strategy to reduce support counting in depth-first search. *SPAM* [5] adopts a vertical bitmap representation, and can mine longer sequential patterns in the cost of more space. *FreeSpan* [45] first finds frequent itemsets and uses frequent itemsets to assemble sequential patterns.

2.4 Mining Sequential Patterns with Constraints

Although efficient algorithms have been proposed, mining a large amount of sequential patterns from large sequence databases is inherently a computationally expensive task. If we can focus on only those sequential patterns of interest to users, we may be able to avoid a lot of computation cost caused by those uninteresting patterns. This opens a new opportunity for performance improvement: "Can we improve the efficiency of sequential pattern mining by focusing only on interesting patterns?"

For effectiveness and efficiency considerations, constraints are essential in many data mining applications. Consider the following example. To characterize a new disease, researchers may want to find sequential patterns about symptoms, such as "*finding patterns with constraint of* 2-7 *days of cough followed by fever in the range of* 37.5-39C *for* 2-5 *days with average temperature of* $38 \pm 0.2C$, *and all these symptoms appear within a period of 2 weeks.*" A pattern found could be "*cough 5 days and fever 4 days with strong headache.*" This mining query contains a few constraints, involving sequences containing certain constants, and with average functions, etc.

In the context of constraint-based sequential pattern mining, Srikant and Agrawal [101] generalized the scope of sequential pattern mining [3] to include time constraints, sliding time windows, and user-defined taxonomy. Mining frequent episodes in a sequence of events studied by Mannila et al. [76] can also be viewed as a constrained mining problem, since episodes are essentially constraints on events in the form of acyclic graphs. Garofalakis et al. [34] proposed regular expressions as constraints for sequential pattern mining and developed a family of *SPIRIT* algorithms; members in the family achieve various degrees of constraint enforcement. The algorithms use relaxed constraints with nice properties (like anti-monotonicity) to filter out some unpromising patterns/candidates in their early stage. Pei et al. [89] proposed a systematic category of constraints and the pattern-growth methods to tackle the constraints.

2.4.1 Categories of Constraints

A *constraint* C for sequential pattern mining is a boolean function $C(\alpha)$ on sequence patterns α. The *problem of constraint-based sequential pattern mining* is to find the complete set of sequential patterns satisfying a given constraint C.

Constraints can be examined and characterized from different points of view. Below we examine them first from the application point of view and then from the constraint-pushing point of view, and build up linkages between the two.

Constraints in Applications

From the application point of view, we present the following seven categories of constraints based on the semantics and the forms of the constraints. Although these are by no means complete, they cover many interesting constraints in applications.

An *item constraint* specifies subset of items that should or should not be present in the patterns. It is in the form of

$$C_{item}(\alpha) \equiv (\varphi i : 1 \leqslant i \leqslant len(\alpha),\ \alpha[i]\ \theta\ V),$$

or

$$C_{item}(\alpha) \equiv (\varphi i : 1 \leqslant i \leqslant len(\alpha),\ \alpha[i] \cap V \neq \emptyset),$$

where V is a subset of items, $\varphi \in \{\forall, \exists\}$ and $\theta \in \{\subseteq, \not\subseteq, \supseteq, \not\supseteq, \in, \notin\}$. For the sake of brevity, we omit the strict operators (e.g., $\subset, \supset$) in our discussion here. However, the same principles can be applied to them.

For example, when mining sequential patterns over a web log, a user may be interested in only patterns about visits to online bookstores. Let B denote the set of online bookstores. The corresponding item constraint is

$$C_{bookstore}(\alpha) \equiv (\forall i : 1 \leqslant i \leqslant len(\alpha), \alpha[i] \subseteq B)$$

A *length constraint* specifies a requirement on the length of the patterns, where the length can be either the number of occurrences of items or the number of transactions. Length constraints can also be specified as the number of distinct items, or even the maximal number of items per transaction.

For example, a user may want to find only long patterns (e.g., patterns consisting of at least 50 transactions) in bio-sequence analysis. Such a requirement can be expressed by a length constraint

$$C_{len}(\alpha) \equiv (len(\alpha) \geqslant 50)$$

A *super-pattern constraint* is in the form of

$$C_{pat}(\alpha) \equiv (\exists \gamma \in P\ s.t.\ \gamma \sqsubseteq \alpha),$$

where P is a given set of patterns, i.e., to find patterns that contain a particular set of patterns as sub-patterns.

For example, an analyst might want to find sequential patterns that first buy a PC and then buy a digital camera. The constraint can be expressed as

$$C_{pat}(\alpha) \equiv (PC)(digital\_camera) \sqsubseteq \alpha.$$

An *aggregate constraint* is a constraint on an aggregate of items in a pattern, where the aggregate function can be `sum`, `avg`, `max`, `min`, standard deviation, etc.

For example, a marketing analyst may want sequential patterns α where the average price of all the items in α is over $100.

A *regular expression constraint* C_{RE} is a constraint specified as a regular expression over the set of items using the established set of regular expression operators, such as disjunction and Kleene closure. A sequential pattern satisfies C_{RE} if and only if the pattern is accepted by its equivalent deterministic finite automata.

For example, to find sequential patterns about a Web click stream starting from Yahoo's home page and reaching hotels in New York city, one may use the following regular expression constraint

$$\textit{Travel}\,(\textit{New York} \mid \textit{New York City}\,)\;(\;\textit{Hotels} \mid \textit{Hotels and Motels} \mid \textit{Lodging}\,),$$

where "|" stands for disjunction. The concept of regular expression constraint for sequential pattern mining was first proposed in [34].

In some applications, one may want to have constraints on the duration of the patterns.

A *duration constraint* is defined only in sequence databases where each transaction in every sequence has a time-stamp. (It is clear that the principle can be readily applied to all sequences, by viewing sequence positions as the timestamps.) It requires that the sequential patterns in the sequence database must have the property such that the time-stamp difference between the first and the last transactions in a sequential pattern must be longer or shorter than a given period.

Formally, a duration constraint is in the form of

$$C_{dur} \equiv Dur(\alpha)\ \theta\ \Delta t,$$

where $\theta \in \{\leqslant, \geqslant\}$ and Δt is a given integer. A sequence α satisfies the constraint if and only if

$$|\{\beta \in SDB \mid \exists 1 \leqslant i_1 < \cdots < i_{len(\alpha)} \leqslant len(\beta)$$

such that

$$\alpha[1] \sqsubseteq \beta[i_1], \ldots, \alpha[len(\alpha)] \sqsubseteq \beta[i_{len(\alpha)}],$$

and

$$(\beta[i_{len(\alpha)}].time - \beta[i_1].time)\ \theta\ \Delta t\}| \geqslant min\_sup$$

In some other applications, the gap between adjacent transactions in a pattern may be important.

A *gap constraint* set is defined only in sequence databases where each transaction in every sequence has a timestamp. It requires that the sequential patterns in the sequence database must have the property such that the timestamp difference between every two adjacent transactions must be longer or shorter than a given gap.

Formally, a gap constraint is in the form of

$$C_{gap} \equiv Gap(\alpha)\ \theta\ \Delta t,$$

where $\theta \in \{\leqslant, \geqslant\}$ and Δt is a given integer. A sequence α satisfies the constraint if and only if

$$|\{\beta \in SDB | \exists 1 \leqslant i_1 < \cdots < i_{len(\alpha)} \leqslant len(\beta)$$

such that

$$\alpha[1] \sqsubseteq \beta[i_1], \ldots, \alpha[len(\alpha)] \sqsubseteq \beta[i_{len(\alpha)}],$$

and for all $1 < j \leqslant len(\alpha)$,

$$(\beta[i_j].time - \beta[i_{j-1}].time)\ \theta\ \Delta t\}| \geqslant min\_sup$$

Among the constraints listed above, duration constraints and gap constraints are *support-related*, that is, they are applied to confine how a sequence matches a pattern. To find whether a sequential pattern satisfies these constraints, one needs to examine the sequence databases. For other constraints, whether the constraint is satisfied can be determined by the frequent patterns themselves without referring to the support counting process.

Characterization of Constraints

In the recent studies of constrained frequent pattern mining [81, 86, 87, 5], constraints are characterized based on the notion of monotonicity, anti-monotonicity, succinctness, and whether they can be transformed into these categories if they do not belong to them. This has become a classical framework for constraint-based frequent pattern mining. This framework can also be extended to constrained sequential pattern mining.

A constraint C_A is *anti-monotonic* if a sequence α satisfying C_A implies that every non-empty subsequence of α also satisfies C_A. A constraint C_M is *monotonic* if a sequence α satisfies C_M implies that every super-sequence of α also satisfies C_M. The basic idea behind *succinct constraint* is that, with such a constraint, one can explicitly and precisely generate all the sets of items satisfying the constraint without recourse to a generate-everything-and-test approach. A succinct constraint is specified using a precise "formula".

According to the "formula", one can generate all the patterns satisfying a succinct constraint. There is no need to iteratively check the constraint in the mining process. Limited by space, we omit the formal definitions here.

For example, length constraint $C_{len}(\alpha) \equiv len(\alpha) \leqslant 10$ and duration constraint $Dur(\alpha) \leqslant 30$ are anti-monotonic, while super-pattern constraint and the duration constraint $Dur(\alpha) \geqslant 30$ are monotonic. It is easy to show that item constraints, length constraints and super-pattern constraints are all succinct.

Based on the above definition, the anti-monotonic, monotonic and succinct characteristics of some commonly used constraints for sequential pattern mining are shown in Table 2.4.

	Constraint	Anti-mono	Mono	Succ
Item	$C_{item}(\alpha) \equiv (\forall i : 1 \leqslant i \leqslant len(\alpha), \alpha[i]\theta V)$ $(\theta \in \{\subseteq, \supseteq\})$	Yes	No	Yes
	$C_{item}(\alpha) \equiv (\forall i : 1 \leqslant i \leqslant len(\alpha), \alpha[i] \cap V \neq \emptyset)$	Yes	No	Yes
	$C_{item}(\alpha) \equiv (\exists i : 1 \leqslant i \leqslant len(\alpha), \alpha[i]\theta V)$ $(\theta \in \{\subseteq, \supseteq\})$	No	Yes	Yes
	$C_{item}(\alpha) \equiv (\exists i : 1 \leqslant i \leqslant len(\alpha), \alpha[i] \cap V \neq \emptyset)$	No	Yes	Yes
Length	$len(\alpha) \leqslant l$	Yes	No	Yes
	$len(\alpha) \geqslant l$	No	Yes	Yes
Super-pattern	$C_{pat}(\alpha) \equiv (\exists \gamma \in P \text{ s.t. } \gamma \sqsubseteq \alpha)$	No	Yes	Yes
Simple aggregates	$max(\alpha) \leqslant v), min(\alpha) \geqslant v$	Yes	No	Yes
	$max(\alpha) \geqslant v), min(\alpha) \leqslant v$	No	Yes	Yes
	$sum(\alpha) \leqslant v$ (with non-negative values)	Yes	No	No
	$sum(\alpha) \geqslant v$ (with non-negative values)	No	Yes	No
Tough aggregates	g_sum: $sum(\alpha)\ \theta\ v, \theta \in \{\leqslant, \geqslant\}$ (†)	No	No	No
	average: $avg(\alpha)\ \theta\ v$	No	No	No
RE	(Regular Expression) §	No	No	No
Duration	$Dur(\alpha) \leqslant \Delta t$	Yes	No	No
	$Dur(\alpha) \geqslant \Delta t$	No	Yes	No
Gap	$Gap(\alpha)\ \theta\ \Delta t\ (\theta \in \{\leqslant, \geqslant\})$	Yes	No	No

Table 2.4. Characterization of commonly used constraints. († with positive and negative values. § in general, a regular expression (RE) constraint is not necessarily anti-monotonic, monotonic, or succinct, though there are cases that are anti-monotonic, monotonic, or succinct. For example, constraint "*" (every pattern satisfies this constraint) is anti-monotonic, monotonic and succinct.)

From Table 2.4, one can see that the classical constraint-pushing framework [81] based on anti-monotonicity, monotonicity, and succinctness can be applied to many constraints. Thus the corresponding constraint-pushing strategy can be integrated easily into any sequential pattern mining algorithms, such as *GSP*, *SPADE*, and *PrefixSpan*. However, some important classes of constraints, such as RE (regular expressions), average(i.e., $avg(\alpha)\ \theta\ v$, where $\theta \in \{\leqslant, \geqslant\}$), and g_sum (i,e., *sum* of positive and negative values), do not fit into this framework.

This problem, with respect to commonly used regular expression constraints, has been pointed out by Garofalakis et al. [34]. They provided a solution of a set of four *SPIRIT* algorithms, each pushing a stronger relaxation of regular expression constraint $\mathcal{R}$ than its predecessor in the pattern mining loop. The first *SPIRIT* algorithm $SPIRIT(N)$ ("N" for "Naive") only prunes candidate sequences containing elements that do not appear in $\mathcal{R}$. The second one, $SPIRIT(L)$ ("L" for "Legal"), requires every candidate sequence to be *legal* with respect to some state of $A_{\mathcal{R}}$. The third, $SPIRIT(V)$ ("V" for "Valid"), filters out candidate sequences that are not *valid* with respect to any state of $A_{\mathcal{R}}$. The fourth, $SPIRIT(R)$ ("R" for "Regular"), pushes $\mathcal{R}$ all the way inside the mining process by counting support only for valid candidate sequences. $SPIRIT(R)$ looks most promising in constraint pushing. However, when the RE constraint is not highly selective, the experiments reported in [34] show that the number of candidates generated by $SPIRIT(R)$ explodes and the algorithm fails to even complete execution for certain cases (run out of virtual memory). Thus finally, $SPIRIT(V)$ was recommended as the overall winner.

2.4.2 Mining Sequential Patterns with Prefix-Monotone Constraints

As shown in Table 2.4, some important and popularly used constraints do not have the anti-monotonic or monotonic property. Instead of considering such constraints as tough ones and finding different kinds of relaxation or patching tricks to "squeeze" them into the *Apriori* framework, we explore an intuitive way by adopting the pattern-growth sequential pattern mining framework so that such *ugly* constraints can be naturally pushed deeply into the mining process. A tough constraint, such as a regular expression constraint, matches naturally with prefix-based expansion and can be pushed deeply into the subsequence expansion-based mining process. Moreover, the classical anti-monotonic, monotonic, and succinct constraints can be easily adapted to this evaluation framework as well.

We first present a *prefix-monotone property* for constraints and show that most of the constraints discussed so far are prefix-monotone. Then, we develop an efficient mining algorithm to push such constraints into sequential pattern mining.

2.4.3 Prefix-Monotone Property

Let R be an order of items in a sequence database. Since the item ordering in the same transaction is irrelevant to sequential patterns, it is convenient to assume that all items in a transaction are written with respect to the order R.

For example, let R be the alphabetical order. A sequence should be written in the form of $(ade)(bc)$ instead of $(dae)(cb)$. The fact that item x precedes item y in order R is denoted by $x \prec y$.

A constraint C_{pa} is called *prefix anti-monotonic* if for each sequence α satisfying the constraint, it is the case that every prefix of α also satisfies the constraint. A constraint C_{pm} is called *prefix monotonic* if for each sequence α satisfying the constraint, every sequence having α as a prefix also satisfies the constraint. A constraint is called *prefix-monotone* if it is prefix anti-monotonic or prefix monotonic.

It is easy to see that an anti-monotonic constraint is prefix anti-monotonic. A monotonic constraint is also prefix monotonic. For example, the length constraint $len(\alpha) \leqslant 10$ is anti-monotonic. It must also be prefix anti-monotonic. This is because if the length of a sequence α is no more than 10, the length of every prefix of α must be no more than 10 as well. Similarly, $len(\alpha) \geqslant 10$ is prefix monotonic since if the length of any prefix of α is no less than 10, α must be no less than 10 as well.

A succinct constraint is not necessarily prefix anti-monotonic or prefix monotonic. However, since succinct constraints can be pushed deep directly into the mining process (no matter which sequential pattern mining method is applied), the pushing of such constraints will not be analyzed further here.

Now, let us examine the regular expression constraints. A well-known result from the formal language theory is that for every regular expression E, there exists a deterministic finite automata M_E such that M_E accepts exactly the language generated by E.

Given a regular expression E, let M_E be the corresponding (deterministic finite) automata. Let α be a sequence. α is called *legal* with respect to E if a state in M_E can be reached following α. From a regular expression constraint E, we can derive a constraint L_E such that a sequence α satisfies L_E if and only if α is legal with respect to E. Clearly, for each sequence α satisfying the regular expression constraint E, every prefix of α must be legal with respect to E. Furthermore, for each sequence β legal with respect to E, every prefix of β must also be legal with respect to E. Thus, for a given regular expression constraint E, let L_E be the constraint on legal prefix with respect to E. Constraint L_E is prefix anti-monotonic.

So far, we have shown that all the commonly used constraints discussed in Section 2.4.1, except for g_sum and average, have prefix-monotone property.

The prefix-monotone property covers more commonly used constraints than traditional anti-monotonic and monotonic properties, since the prefix-monotone property is weaker than the anti-monotone and monotone properties. All anti-monotonic or monotonic constraints are prefix-monotonic, but the reverse direction is not true.

Often, mining with a weaker constraint may lead to the less efficiency. Thus, one may wonder whether mining with the prefix-monotone property is less efficient than mining using the classical anti-monotonicity-based *Apriori* methods. It turns out that using the prefix-monotone property is more efficient since the pattern-growth, depth-first search methods are more efficient than the breadth-first search methods.

2.4.4 Pushing Prefix-Monotone Constraints into Sequential Pattern Mining

Now, let us examine an example of constraint pushing.

Example 2.12 (Pushing prefix-monotone constraints). Consider the sequence database SDB in Table 2.5 and the task of mining sequential patterns with a regular expression constraint $C = a * \{bb|(bc)d|dd\}$ and support threshold $min\_sup = 2$. The mining can be conducted in the following steps.

Sequence_id	Sequence
10	$a(bc)e$
20	$e(ab)(bc)dd$
30	$c(aef)(abc)dd$
40	$addcb$

Table 2.5. Sequence database SDB.

1. *Find length-*1 *patterns and remove irrelevant sequences.* Similar to sequential pattern mining without constraint C, one needs to scan SDB once, which identifies patterns a, b, c, d, and e as length-1 patterns. Infrequent items, such as f, are removed. Also, in the same scan, the sequences that contain no subsequence satisfying the constraint, such as the first sequence, $a(bc)e$, should be removed.
2. *Divide the set of sequential patterns into subsets without overlap.* Without considering constraint C, the complete set of sequential patterns should be divided into five subsets without overlap according to the set of length-1 sequential patterns: (1) those with prefix a; (2) those with prefix b; $\ldots$; and (5) those with prefix e. However, since only patterns with prefix a may satisfy the constraint C, i.e. only a is legal with respect to constraint C, the other four subsets of patterns are pruned.
3. *Construct a-projected database and mine it.* Only the sequences in SDB containing item a and satisfying constraint C should be projected. The a-projected database, $SDB|_a = \{(\_b)(bc)dd, (\_e)(abc)dd, ddcb\}$. Notice that $e(ab)(bc)dd$ is projected as $(\_b)(bc)dd$, where symbol "_" in the first transaction indicates that it is in the same transaction with a.
During the construction of the a-projected database, we also find locally frequent items: (1) b can be inserted into the same transaction with a to form a longer frequent prefix (ab), and (2) b, c and d can be concatenated to a to form longer frequent prefixes, i.e., ab, ac and ad. Locally infrequent items, such as e, should be ignored in the remaining mining of this projected database.

Then the set of patterns with prefix a can be further divided into five subsets without overlap: (1) pattern a itself; (2) those with prefix (ab); (3) those with prefix ab; (4) those with prefix ac; and (5) those with prefix ad. With the existence of constraint C, pattern a fails C and thus is discarded; and (ab) is illegal with respect to constraint C, so the second subset of patterns is pruned. The remaining subsets of patterns should be explored one by one.

4. *Mine subsets recursively.* To mine patterns having ab as a prefix, we form the ab-projected database $SDB|_{ab} = \{(\_c)dd, (\_c)dd\}$. By recursively mining the projected database, we identify sequential pattern $a(bc)d$ which satisfies the constraint.
To mine patterns with prefix ac, we form ac-projected database $SDB|_{ac} = \{dd, dd, b\}$. Every sequence in the projected database contains no subsequence satisfying the constraint. Thus, the search within $TDB|_{ac}$ can be pruned. In other words, we will never search any projected database which does not potentially support patterns satisfying the constraint.
Similarly, we search the ad-projected database and find add is a sequential pattern satisfying the constraint.
In summary, during the recursive mining, if the prefix itself is a pattern satisfying the constraint, it should be an output. The prefixes legal with respect to the constraint should be grown and mined recursively. The process terminates when there is no local frequent item or there is no legal prefix. It results in two final patterns: $\{a(bc)d, add\}$. ∎

Let's verify the correctness and completeness of the mining process described in Example 2.12. As shown in the example, if $x_1, \ldots, x_n$ are the complete set of length-1 patterns, the complete set of sequential patterns can be divided into n subsets without overlap and the mining can also be reduced to the mining of n projected databases. Such a divide-and-conquer strategy can be applied recursively. This is the depth-first search in the *PrefixSpan* algorithm.

We can remove a sequence that does not satisfy the constraint according to the following rule.

Lemma 2.13. *Let the given constraint be C.*

1. *In a sequence database SDB, if a sequence α does not contain any subsequence satisfying C, then the set of sequential patterns satisfying C in SDB is identical to the set of sequential patterns satisfying C in $SDB - \{\alpha\}$.*
2. *Let $\alpha \cdot \beta$ be a sequence in the α-projected database $SDB|_\alpha$. If there exists no a subsequence $\alpha \cdot \gamma \sqsubseteq \alpha \cdot \beta$ satisfying C, then the set of sequential patterns satisfying C in $SDB|_\alpha$ is identical to the set of sequential patterns satisfying C in $SDB|_\alpha - \{\alpha \cdot \beta\}$.*

Proof. We prove the first claim. The second claim can be proved similarly.

Since $SDB - \{\alpha\} \subset SDB$, every sequential pattern in $SDB - \{\alpha\}$ is also a sequential pattern in SDB. Thus, every sequential pattern in $SDB - \{\alpha\}$ satisfying C is also a sequential pattern in SDB satisfying C. Moreover, for any sequence β that is not a subsequence of α, the support of β in $SDB - \{\alpha\}$ is identical to the support of β in SDB.

Suppose there exists a sequential pattern γ satisfying C in SDB but not in $SDB - \{\alpha\}$. γ must be a subsequence of α. That contradicts the assumption that α does not contain any subsequence satisfying C. Thus, every sequential pattern satisfying C in SDB is also a sequential pattern satisfying C in $SDB - \{\alpha\}$. The claim is proved. ■

With a prefix-monotone constraint, one only needs to search in the projected databases having prefixes potentially satisfying the constraint, as suggested in the following lemma.

Lemma 2.14. *Given a prefix-monotone constraint C. Let α be a sequential pattern.*

1. *When C is prefix anti-monotonic, if $C(\alpha) = false$, then there exist no sequential patterns that have α as a prefix and also satisfy constraint C.*
2. *When C is prefix monotonic, if $C(\alpha) = true$, then every sequential pattern having α as a prefix satisfies C.*
3. *When C is a regular expression constraint, if α is illegal with respect to C, then there exist no sequential patterns that have α as a prefix and also satisfy constraint C.*

Proof. The lemma follows the prefix-monotone property immediately. In case 2, constraint testing can be waived for mining in $SDB|_\alpha$. ■

Based on the above discussion, we have the constrained sequential pattern mining algorithm *Prefix-growth* as given in Figure 2.4.

Let us analyze the efficiency of the *Prefix-growth* algorithm.

First, Algorithm *Prefix-growth* takes *PrefixSpan* as the basic sequential pattern mining algorithm and pushes prefix-monotone constraints deeply into the *PrefixSpan* mining process. The performance study in [88] shows that *PrefixSpan* outperforms *GSP*, owing to the following factors.

- *PrefixSpan* adopts a prefix growth and database projection framework: for each frequent prefix subsequence, only its corresponding suffix subsequences need to be projected and examined without candidate generation.
- *PrefixSpan* applies a divide-and-conquer strategy so that sequential patterns are grown by exploring only local frequent patterns in each projected database.
- *PrefixSpan* explores further optimizations, including a *pseudo-projection* technique when the projected database and its associated pseudo-projection processing structure fits in main memory, etc.

Input: A sequence database SDB, support threshold min_sup, and prefix-monotone constraint C;
Output: The complete set of sequential patterns satisfying C;
Method:
call $prefix\_growth(\emptyset, SDB)$.
Function $prefix\_growth(\alpha, SDB|_\alpha)$
// α: prefix; $SDB|_\alpha$: the α-projected database

1. Let l be the length of α. Scan $SDB|_\alpha$ once, find length-$(l+1)$ frequent prefix in $SDB|_\alpha$, and remove infrequent items and useless sequences;
2. for each length-$(l+1)$ frequent prefix α' potentially satisfying the constraint C (Lemma 2.14) do
   a) if α' satisfies C, then output α' as a pattern;
   b) form $SDB|_{\alpha'}$;
   c) call $prefix\_growth(\alpha', SDB|_{\alpha'})$

Fig. 2.4. The *Prefix-growth* algorithm.

Second, *Prefix-growth* handles a broader scope of constraints than anti-monotonicity and monotonicity. A typical such example is regular expression constraints, which are difficult to address using an *Apriori*-based method, as shown in *SPIRIT*. By *Prefix-growth*, such constraints can be naturally pushed deeply into the mining process.

Both *Prefix-growth* and *SPIRIT* [34] push regular expression constraints by relaxing the constraint to achieve some nice property facilitating the constraint-based pruning. In particular, *SPIRIT(V)* requires every candidate sequence to be valid with respect to some state of the automata $\mathcal{A}_{\mathcal{R}}$, which shares a similar idea with *Prefix-growth*. However, the *SPIRIT* methods adopt the candidate-generation-and-test framework, which is more costly than the pattern growth methods. Moreover, the *SPIRIT* methods are dedicated to pushing regular expression constraints, while *Prefix-growth* is capable of pushing many constraints more than regular expression ones. For example, anti-monotonic or monotonic constraints that are not regular expression constraints, such as super-pattern constraints and some aggregate constraints in Table 2.4, can be consistently pushed in *Prefix-growth*, but cannot be handled by SPIRIT. As shown in the experimental results, *Prefix-growth* outperforms *SPIRIT* in pushing regular expression constraints.

Third, constraint checking by Lemma 2.13 further shrinks projected databases effectively, due to its removal of useless sequences with respect to a given constraint during the prefix growth. This ensures that search is pursued in promising space only. Since many irrelevant sequences can be pruned in large databases, the projected databases keep shrinking quickly.

One may wonder whether the *Apriori*-based methods, such as *GSP* and *SPADE*, can do similar prefix-based pruning using prefix-monotone constraints. Taking a non-anti-monotonic regular expression constraint as an example, for *Apriori*-based methods, a pattern whose prefix fails a constraint cannot be pruned since inserting more items/transactions to the pattern at *some other* positions may still lead to a valid pattern. However, by exploring prefix-monotone constraints, *Prefix-growth* puts stronger restrictions on the possible subsequences to grow and thus prunes search space more effectively.

In summary, *although prefix-monotone property is weaker than* Apriori *property, since* Prefix-growth *uses a different methodology for the mining, it still achieves better performance than* Apriori-*based methods.*

2.4.5 Handling Tough Aggregate Constraints by *Prefix-growth*

Besides regular expression constraints, one may wonder whether *Prefix-growth* can effectively handle the two tough aggregate constraints in Table 2.4, average and g_sum? Both constraints are neither anti-monotonic nor monotonic. Even worse, they are not prefix-monotone! Let's examine such an example.

Example 2.15. Let us mine sequential patterns with constraint $C \equiv avg(\alpha) \leqslant 25$ in a sequence database SDB as shown in Table 2.6, with support threshold = 2. The four items in the database are of values 10, 20, 30 and 50, respectively. For convenience, the item values are used as Ids of items.

Sequence_id	Sequence
10	50 10 20 20
20	30 50 20
30	50 10 20 10 10
40	30 20 10

Table 2.6. Another sequence database SDB.

Constraint C cannot be directly pushed into the *PrefixSpan* mining process. For example $\alpha = 50$ cannot be discarded even $avg(\alpha) \not\leqslant 25$, since by appending more elements to α, we may have $\alpha' = 50\ 10\ 20\ 10$ and $avg(\alpha') \leqslant 25$. Also, one can easily verify that C is not prefix-monotone.

In [87], a technique was developed to push *convertible* constraints, like $avg(X) \geqslant 25$, into frequent itemset mining on *transactional databases*. The general idea is to use a proper order of frequent items, like value descending order for constraint $avg(X) \geqslant v$, such that the list of frequent items according to the order has a nice anti-monotonic or monotonic property.

Can we apply the technique in [87] to tackle the aggregate constraints for sequential pattern mining? Unfortunately, the answer is negative. For every sequence, a temporal order has been pre-composed and we do not have the freedom to re-arrange the items in sequences. The trick of simple ordering does not work well here.

Thus, new constraint pushing methods should be explored. ∎

Let's examine how to push constraint "$avg(\alpha) \leqslant v$" deep into the *Prefix-growth* mining process.

Value-ascending order over the set of items should be used to determine the order of projected databases to be processed. An item i is called a small item if its value $i.value \leqslant v$, otherwise, it is called a big item.

In the first scan of a (projected) database, unpromising big items in sequences should be removed according to the following two rules.

Lemma 2.16 (Pruning unpromising sequences). *In mining sequential patterns with constraint $avg(\alpha) \leqslant v$, for a sequence α, let n be the number of instances of small items and s be the sum of them. If there are multiple instances of one small item, the value of that item should be counted multiple times. For any big item x in α such that $\frac{s+x.value}{n+1} > v$, there exists no subsequence $\beta \sqsubseteq \alpha$ that contains x and $avg(\beta) \leqslant v$.*

Proof. Consider any $\beta \sqsubseteq \alpha$ such that x is in β. Clearly, the occurrences of small items in β are a subset of the occurrences of small items in α. Thus, $avg(\beta) \geqslant \frac{s+x.value}{n+1} > v$. ∎

The big items identified by Lemma 2.16 are called *unpromising*. Removal of unpromising big items do not imply that we will miss any sequential patterns satisfying the constraint. Instead, it helps us to shrink the sequence database and it facilitates the mining.

Similarly, unpromising sequence pruning rule can be recursively applied in an α-projected database. For a projection $\gamma = \beta/\alpha$, let n be the number of instances of small items appearing in γ but not in α and s be the sum of them. A big item x in α is unpromising and should be removed if $(s + sum(\alpha) + x.value)/(n + \#items(\alpha) + 1)$ violates the constraint. Here, function $\#items(\alpha)$ returns the number of instances of items in sequence α.

Moreover, an item marking method can be developed to mark and further prune some unpromising items as follows. In the α-projected database[3], when a pattern β is found where the first item following α is a small item, we check whether that small item can be replaced by a big item x frequent in the projected database and still can get average value satisfying the constraint. If so, prefix $\alpha \cdot x$ is marked *promising* and does not need to be checked and marked again in this projected database. When all patterns with some small item as the first one following α have been found, for the prefixes with a big item x following α having not been marked, $\alpha \cdot x$ as well as the projected databases can be pruned if $\alpha \cdot x$ violates the constraint. We call this the **unpromising pattern pruning rule**.

The rationale of this rule is as follows. For a big item x, if $\alpha \cdot x$ violates the constraint but $\alpha \cdot x \cdot \beta$ is a sequential pattern satisfying the constraint, then there must be some $\beta' \sqsubseteq \beta$ such that β' starts with a small item and $\alpha \cdot \beta'$ is a sequential pattern satisfying the constraint.

---

[3] The whole database SDB can be regarded as $SDB|$.

We elaborate on the rules and the mining procedure using the SDB in Example 2.15 as follows.

Example 2.17. Let us consider mining SDB in Table 2.6 with constraint $C \equiv avg(\alpha) \leqslant 25$.

In the first scan of SDB, we remove the unpromising big items in sequences by applying the *unpromising sequence pruning rule.* For example, in the second sequence, 20 is the only small item and $\frac{20+50}{2} = 35 > 25$. This sequence cannot support any sequential patterns having item 50 and satisfying the constraint C. Thus, item 50 in the second sequence should be moved.

In the same database scan, we also find length-1 patterns, 10, 20, 30 and 50. The set of patterns can be partitioned into four subsets without overlap: (1) those with prefix 10; (2) those with prefix 20; (3) those with prefix 30; and (4) those with prefix 50. These subsets of patterns should be explored one by one in this order.

1. The set of patterns with prefix 10 can be found by constructing 10-projected database and mining it recursively. The items in 10-projected database are small ones, so all patterns in it have average no greater than 25 and thus satisfy the constraint. There are two patterns there: 10 and 10 20.
   When pattern 10 is found, it can be regarded as a small item 10 following the empty prefix. Thus, we apply the *unpromising pattern pruning rule* to mark and prune patterns. Prefix 30 is marked as *promising*, since $avg(30 \cdot 10) = 20 < 25$. Prefix 30 will not be checked against any other pattern after it is marked.
   None of the patterns with prefix 10 can be used to mark prefix 50.
2. Similarly, we can find patterns with prefix 20 by constructing and mining 20-projected database. They are 20 and 20 10. None of the patterns with prefix 20 can be used to mark prefix 50.
3. 30 is a big item and prefix 30 violates the constraint. For patterns with prefix 30, since prefix 30 is marked, we need to construct 30-projected database and mine it. Pattern 30 20 is found.
4. Prefix 50 has not been marked. According to the unpromising pattern pruning rule, no pattern with prefix 50 can satisfy the constraint. We do not need to construct or mine 50-projected database. ■

Constraint $avg(\alpha) \geqslant v$ is dual with respect to constraint $avg(\alpha) \leqslant v$. Therefore, we can prune unpromising (small) items and patterns similarly. Moreover, with the same idea, constraint $sum(\alpha)\ \theta\ v$ (where $\theta \in \{\leqslant, \geqslant\}$, and items can be with non-negative and negative values) can also be pushed deeply into *Prefix-growth* mining process. This is left as an exercise for interested readers.

Thus, although the two concrete rules discussed above are specific for constraint $avg(\alpha) \leqslant v$, the idea is general and can be applied to some other aggregate constraints. The central point is that we can prune unpromising items and patterns as the depth-first search goes deep. In summary, with minor revision, *Prefix-growth* can be extended to handle some tough aggregate constraints without prefix-monotone property. With such extensions, all established advantages of *Prefix-growth* still retain and the pruning is still sharp. This is also verified by experimental results.

Pushing Multiple Constraints

We have studied the push of individual constraints into sequential pattern mining. Can we push multiple constraints deep into sequential pattern mining process?

Multiple constraints in a mining query may belong to the same category (e.g., all are anti-monotonic) or to different ones. Moreover, different constraints may be on different properties of items/transactions (e.g., some could be on item price, while others could be on timestamps, etc.).

A constraint in the form of conjunctions and/or disjunction of prefix-monotone constraints still can be pushed deep into a *Prefix-growth* mining process. We only need to keep track of which sub-constraints have been satisfied/violated. Based on that, whether the whole constraint is satisfied and whether further recursive mining of projected databases is needed can be determined. The details will be left to interested readers as an exercise.

For constraints involving aggregates $avg()$ and $sum()$ (where items can be with non-negative and negative values), *Prefix-growth* uses a global order over all items to push them into the mining process. However, when a constraint involves more than one of such aggregates, and the orders required by these sub-constraints are conflicting, some coordination is needed. The general philosophy is to conduct a cost analysis to determine how to combine multiple order-consistent constraints and how to select a sharper constraint among order-conflicting ones. The details will be left to interested readers as an exercise.

2.5 Mining Closed Sequential Patterns

The complete set of sequential patterns in a sequence database may contain redundant information. In this section, we identify such redundancy and introduce the notion of closed sequential patterns. Moreover, we also discuss how to mine closed sequential patterns efficiently.

2.5.1 Closed Sequential Patterns

Example 2.18 (Redundancy in sequential patterns). Consider a sequence database SDB in Table 2.7. Suppose the support threshold is *min_support*= 1. Let us consider the sequential patterns in the database.

Sequence-id	Sequence
10	$a_1 \cdots a_{50}$
20	$a_1 \cdots a_{100}$

Table 2.7. A database of long sequences.

Clearly, any sequence $a_{i_1} \cdots a_{i_k}$ $(1 \leqslant i_1 < \cdots < i_k \leqslant 100)$ is a sequential pattern. The total number of sequential patterns is

$$\sum_{i=1}^{100} \binom{100}{i} = 2^{100} - 1 = 1,267,650,600,228,229,401,496,703,205,376.$$

For a sequential pattern $a_{i_1} \cdots a_{i_k}$, if $i_k \leqslant 50$ then $sup(a_{i_1} \cdots a_{i_k}) = 2$. On the other hand, if $i_k > 50$, then $sup(a_{i_1} \cdots a_{i_k}) = 1$.

In other words, the more than 1.2×10^{30} sequential patterns can be summarized by two patterns, $s_1 = a_1 \cdots a_{50}$ and $s_2 = a_1 \cdots a_{100}$ and their support information, $sup(s_1) = 2$ and $sup(s_2) = 1$. Due to the anti-monotonic property (Theorem 2.3), every subsequence of s_1 has support 2, and every subsequence of s_2 but not a sequence of s_1 has support 1. ∎

The idea elaborated in the above example motivates the proposal of closed sequential patterns [125], which is an extension of the idea of frequent closed itemsets [84].

Definition 2.19 (Closed sequential pattern). *A sequential pattern s is called* closed *if there is no proper super-pattern $s' \sqsupset s$ such that $sup(s) = sup(s')$.* ∎

The set of closed sequential patterns is a lossless compression of the set of all sequential patterns. For a sequence p, whether p is a sequential pattern and its support information can be derived from the set of closed sequential patterns as follows.

- p is not a sequential pattern (that is, $sup(p)$ is less than the minimum support threshold) if and only if there exists no any closed sequential pattern s such that $p \sqsubseteq s$.
- If p is a sequential pattern, then $sup(p) = sup(s)$ where s is a closed sequential pattern such that $p \sqsubseteq s$ and there exists no any other closed sequential pattern s' such that $p \sqsubseteq s' \sqsubset s$.

The correctness of the above rules can be proved using the anti-monotonic property (Theorem 2.3) and the definition of closed sequential patterns. It can also be shown that, in the second item, closed sequential pattern s is unique. We leave the details as an exercise for the interested readers.

2.5.2 Efficiently Mining Closed Sequential Patterns

Given a sequence database and a minimum support threshold, how can we mine the complete set of closed sequential patterns efficiently?

One naïve approach works as follows. First, we can find the complete set of sequential patterns using a sequential pattern mining algorithm such as those discussed before. Then, for each sequential pattern, we test whether it is closed. Only those closed patterns are reported.

However, the naïve method is too costly. First, it has to find all sequential patterns. In Example 2.18, it still has to find the more than 1.2×10^{30} sequential patterns! Second, testing whether sequential patterns are closed is non-trivial. For each sequential pattern, we may have to check it against many other sequential patterns to test whether it is a subpattern of the others and has the same support.

A few efficient algorithms have been proposed to mine closed sequential patterns, such as [109, 114, 125]. Here, we discuss the major ideas in *BIDE* [114], which is a representative method.

BIDE is an extension of *PrefixSpan* for closed sequential pattern mining. Several important techniques are developed to prune the search space so that many non-closed sequential patterns do not need to be searched. One central idea is the so-called BI-directional extension closure checking scheme.

To keep our discussion easy to follow, let us assume that each element contains only one item. Suppose in the depth-first search, a sequential pattern $s = e_1 \cdots e_n$ is found. How can we determine whether s is closed or not? Do we need to search any super-sequences of s that have s as a prefix?

If s is not closed, at least one of the following two cases must happen.

- There exists a sequential pattern $s_1 = e_1 \cdots e_n e'$ such that $sup(s) = sup(s')$. In other words, in every sequence that contains s, e' appears after a prefix that is a super-sequence of s. We call s' a *forward-extension sequence* of s, and e' a *forward-extension event.*
- There exists a sequential pattern $s_1 = e_1 \cdots e_i e' e_{i+1} \cdots e_n$ or $s_1 = e' e_1 \cdots e_n$ such that $sup(s) = sup(s_1)$. We call s_1 a *backward-extension sequence* of s, and e' a *backward-extension event.*

Clearly, a sequential pattern p is closed if and only if there exists neither a forward-extension event nor a backward-extension event. Now, the question is how to find forward- and backward- extension events quickly.

Finding forward-extension events is straightforward. For a sequential pattern p, we can form the p-projected database which has all sequences containing p. Then, we check the suffixes of the sequences. If an item e' appears in every suffix, e' must be a forward-extension event.

Finding backward-extension events is a little bit tricky. An item may appear multiple times in a sequence. Therefore, to check whether there is a backward-extension event between e_i and e_{i+1}, for each sequence containing p, we look at the subsequence between the first occurrence of $e_1 \cdots e_i$ and the last occurrence of $e_{i+1} \cdots e_n$. If there is an event e' appearing in that subsequence of every sequence containing p, then e' is a backward-extension event.

By the above BI-directional extension closure checking, we can determine whether a sequential pattern p is closed when it is generated in the depth-first search process. The only information needed in the checking is the set of sequences containing p. Since projected databases are used in *PrefixSpan*, we only need to extend the projected databases in *PrefixSpan* by including the whole sequences instead of only the suffixes.

The backward-extension events can also be used to prune a search subspace where there exists no closed sequential patterns. Suppose when we search a sequential pattern $p = e_1 \cdots e_n$, and a backward-extension event e' is found, that is, each sequence containing p also contains $q = e_1 \cdots e_i e' e_{i+1} \cdots e_n$. Not only we can immediately determine that p is not closed, but also we can determine that, if a pattern p' has p as a prefix, then p' cannot be closed if p' is formed using subsequences having q in the projected database. To see the reason, consider a pattern $p' = e_1 \cdots e_n e_{n+1} \cdots e_l$ that has p as a prefix, and the matching of $e_{i+1} \cdots e_l$ in the sequences follow with the matching of q. Clearly, every such sequence containing p' also contains $q' = e_1 \cdots e_i e' e_{i+1} \cdots e_n e_{n+1} \cdots e_l$. That is, $sup(q') = sup(p')$. Thus, p' is not closed.

As reported in [114], mining closed sequential patterns not only can reduce the number of patterns substantially, but also can reduce the mining time. Mining dense sequence databases where there are many long sequential patterns becomes feasible by using the closed sequential pattern mining techniques.

2.6 Summary

Sequential pattern mining is an essential task in mining sequence data. In the following chapters, we will see that sequential patterns are used in various advanced analytical tasks on sequence data.

In this chapter, we discussed two fundamental frameworks of sequential pattern mining methods: the *Apriori*-like, breadth-first search methods and the pattern-growth, depth-first search methods. The pattern-growth, depth-first search methods are generally more capable of mining long sequential patterns from large sequence databases.

We also discussed the extensions of the pattern-growth, depth-first search methods to mine sequential patterns with constraints. Different kinds of constraints are examined from both the application point of view and the constraint-pushing point of view. We elaborated on how to handle tough aggregate constraints as well.

Redundancy may exist among sequential patterns in a large database. To tackle the problem, the concept of closed sequential patterns was introduced.

Moreover, we illustrated the major ideas of extending the pattern-growth, depth-first search methods in mining closed sequential patterns efficiently.

3

Classification, Clustering, Features and Distances of Sequence Data

This chapter is concerned with the classification and clustering of sequence data, together with sequence features and sequence distance functions. It is organized as follows:

- Section 3.1 provides a general categorization of sequence classification and sequence clustering tasks. There are three general tasks. Two of those tasks concern whole sequences and will be presented here. The third topic, namely sequence motifs (site/position-based identification and characterization of sequence families), will be presented in Chapter 4. The existence of a third task is due to the facts that (a) positions inside sequences are important, a factor which is not present for non-sequence data, and (b) succinct characterizations of sequence families are desired for identifying future members of the families.
- Section 3.2 discusses sequence features, concerning various feature types and general feature selection criteria. Section 3.3 is about sequence similarity/distance functions. The materials in Sections 3.2 and 3.3 will be useful not only for classification and clustering, but also for other topics (such as identification and characterization of sequence families).
- Section 3.4 discusses sequence classification. Several popular sequence classification algorithms are presented and a brief discussion on the evaluation of classifiers and classification algorithms is given.
- Section 3.5 discusses sequence clustering; it includes several popular sequence clustering approaches and a brief discussion on clustering quality analysis.

3.1 Three Tasks on Sequence Classification/Clustering

On sequence data, the following three data mining tasks are related to the general data mining tasks of classification and clustering.

1. The *classification task* is about building classifiers for given classes of sequence data. This is usually achieved by combining some general classification methods, together with appropriate feature selection/construction methods. The classification can be whole sequence based, where one is interested in whether a sequence belongs to a certain class; it can also be site based, where one is interested in (a) whether a sequence contains a site of interest and (b) the actual position of the site in the sequence if it exists.
2. The *clustering task* is about the grouping of given sequences into clusters. This is usually achieved by selecting a general clustering method, and selecting/designing an appropriate distance function on sequences. The distance functions need to address the special properties of sequences and the particular needs of the underlying applications.
3. The third task can be considered as a hybrid of clustering and classification. It is concerned with the identification of certain sequence clusters (usually referred to as sequence families), together with the characterization of the sequence clusters expressed in terms of some patterns or models. The characterizing patterns or models can then be used for classifying sequences into the corresponding families, in addition to providing discriminative information to scientists.
   This task has two flavors. In the first flavor, the criteria for grouping sequences consider the entire sequences under consideration. In the second flavor, the criteria for grouping sequences consider certain windows in which the sequence shows certain discriminative similarities.

For all three tasks, the clusters or sequence families may be disjoint, or may overlap each other. Moreover, the clusters or families do not need to cover all of the given data.

We consider the first two tasks in this chapter, and will consider the third task in the next chapter.

3.2 Sequence Features

3.2.1 Sequence Feature Types

Sequence features[1] can be considered along the following perspectives:

- Explicit versus implicit (constructed): Some features are (based on) patterns that occur in the sequences while others are constructed from properties of the sequences or the objects underlying the sequences. Examples of the former type include various sequence patterns; examples of the latter type include physical/chemical/spatial properties of protein structures associated with the protein sequences.

[1] Feature selection for general, non-sequence data is covered in [67].

- Simple versus complex and rigid versus flexible: Among the sequence patterns that are useful as features, some are very simple and rigid (e.g. k-grams or k-gapped pairs where the relative distance between positions is fixed), while others are complex and flexible (e.g. distinguishing sequence patterns with gap constraints where the relative distance between positions is constrained by a bound).
- Presence/match versus count: A pattern can generate two types of features. In the first, one uses the pattern as a Boolean feature, by simply considering the presence/absence (or match) of the pattern in the sequence (or its corresponding object). In the second, one uses the pattern as a numerical feature, by considering the count of the pattern in the sequence (or its corresponding object). The match/count can be position/site dependent (only matches at certain positions count) or position/site independent (all matches count, regardless of the position). It is also possible to consider inexact matches.

GAATTCTCTGTAACCTCAGAGGTAGATAGA

Fig. 3.1. A DNA sequence

We now describe some of the major types of sequence features; all but the last one are sequence pattern based; the last one is "property"-based. Let $S = s_1...s_n$ be a sequence over an alphabet $\mathcal{A}$. We will use the DNA sequence in Figure 3.1, which has length of 30, to illustrate some of the feature types.

- Each sequence element (together with its position) can be a feature. Typically the sequences are aligned and element-wise comparison is used to determine similarity between sequences.
- k-grams (or k-mers) are features. Let $k \geqslant 1$ be a positive integer. A k-gram is a sequence over $\mathcal{A}$ of length k. For a sequence, the k-grams of interest are those that occur in the sequence.
  For the sequence in Figure 3.1, the 2-gram GA occurs in S but GC does not. The frequency of the 2-gram GA in the sequence is 4.
- Let $k \geqslant 0$ be an integer and x, y be (possibly identical) elements in $\mathcal{A}$. Then the ordered pair (x, y) can be viewed as a *k-gapped pair* (or gapped residue couple [20, 83, 39, 51]), and its *k-gapped occurrence frequency* in S, denoted by $F(xy, k)$ is defined by

$$F(xy, k) = \frac{1}{n-k-1} \Sigma_{i=1}^{n-k-1} O_{x,y}(i, i+k+1)$$

where $O_{x,y}(i, i+k+1) = 1$ if $s_i = x$ and $s_{i+k+1} = y$, and $O_{x,y}(i, i+k+1) = 0$ otherwise. So $F(xy, k)$ is the fraction of matches of x, y in S as a k-gapped pattern over the maximal number of matches possible. (It should

be pointed out that k is the exact fixed gap between the two symbols, and is not used as a maximum gap constraint.) When $k = 0$, the k-gapped pair is equivalent to a 2-gram.
For the DNA sequence in Figure 3.1, $F(GA, 0) = \frac{4}{29}$, $F(GA, 1) = \frac{3}{28}$, $F(GA, 2) = \frac{3}{27}$, etc.

- One may also consider cumulative running counts as features [134]. For each pattern P over $\mathcal{A}$ and each position i, we have a count $C(P, i)$, which is the number of matches of P in $s_1 ... s_i$ (the prefix of S up to s_i). For example, for the DNA sequence of Figure 3.1, we have $C(A, 1) = 0$, $C(A, 3) = 2$, $C(T, 9) = 4$, etc.
- In general, arbitrary sequence patterns (either their presence/absence or counts) can be used as features, including frequent sequence patterns (see Chapter 2) and distinguishing sequence patterns (see Chapter 6).
One can partition the alphabet into equivalence classes (or exchange groups), and then consider k-grams of equivalence classes [123, 116]. This technique has been used for protein sequences, since different amino acids may behave similarly and can be considered as equivalent. The technique can help reduce the number of features when there are too many. For example, we can partition the amino acids of proteins into 6 equivalence classes represented by $\{e_1, e_2, e_3, e_4, e_5, e_6\}$, where

$$\begin{aligned}
e_1 &= \{H, R, K\},\\
e_2 &= \{D, E, N, Q\},\\
e_3 &= \{C\},\\
e_4 &= \{S, T, P, A, G\},\\
e_5 &= \{M, I, L, V\},\\
e_6 &= \{F, Y, W\}.
\end{aligned}$$

(Such equivalence classes represent conservative replacements through evolution; such equivalence among amino acids is derived from PAM [122].) The protein sequence $KALSLLAG$ can be considered as the sequence $e_1e_4e_5e_4e_5e_5e_4e_4$ over the equivalence classes, and hence it can be considered to match the 2-gram of e_4e_5.
- Features can also be property based. For example, for protein sequences, one can use physical/chemical properties of the protein as features. More specifically, one can use the shape, charge, and hydrophobicity at a sequence/structure location as features.

Features in the form of k-grams and k-gapped pairs have been used extensively in biological sequence analysis.

3.2.2 Sequence Feature Selection

Once the set of potential features is determined, there need to be some criteria for selecting the "good" features for potential usage in data mining tasks. Two general criteria can be used:

- Frequency based feature selection: The features with high frequencies, namely those having frequency over a given threshold, are selected. One may use window-based frequency or whole-sequence frequency.
- Discrimination based feature selection: Features with relatively higher frequency at a desired site or in some selected classes than elsewhere are preferred. To find features for identifying a desired site, one prefers features which occur quite frequently around the site than elsewhere. To find features for identifying classes, one prefers features which occur more frequently in one class than in other classes.

Some method is needed to determine patterns' frequency differences as discussed above. This can be done by directly comparing a site against other parts of the sequences, or comparing one class against other classes. In the biological literature, it is often the case that one actually computes the frequency at the desired site or in a class, but uses the so-called "background" probabilities to determine the frequency away from the given site or the given class. The background probabilities can be sampled from large datasets, e.g. all potential sequences in the application.

Chapter 6 gives more details on mining patterns that distinguish a site/class against other parts of the sequences or other classes.

3.3 Distance Functions over Sequences

Sequence distance functions are designed to measure sequence (dis)similarities. Many have been proposed for different sequence characteristics and purposes. After an overview, this section will discuss several major types of sequence distance functions.

3.3.1 Overview on Sequence Distance Functions

A sequence distance is a function d mapping pairs of sequences to non-negative real numbers so that the following four properties are satisfied:

- $d(x, y) > 0$ for sequences x and y such that $x \neq y$,
- $d(x, x) = 0$ for all sequences x,
- $d(x, y) = d(y, x)$ for all sequences x and y,
- $d(x, y) \leqslant d(x, z) + d(z, y)$ for all sequences x, y and z.

Moreover, d is a *normalized* distance function if $0 \leqslant d(x, y) \leqslant 1$ for all sequences x and y. These properties and definitions are special cases of those for general distance functions.

Sequence distance functions can be used for many different purposes, including clustering, finding known sequences similar to a new sequence (to help infer the properties of new sequences), determining the phylogenetic tree (evolution history) of the species, etc.

Roughly speaking, distance functions can be character (alignment) based, feature based, information theoretic (e.g. Kolmogorov complexity) based [64], conditional probability distribution based [126], etc. In the feature based approach, one would first extract features from the sequences, and then compute the distance (e.g. Euclidean or cosine) between the sequences by computing the distance between the feature vectors of the sequences. The character alignment based ones can be local window based or whole sequence based; they can also be edit distances or more general pairwise similarity score based distances. Details on most of these approaches are given below.

3.3.2 Edit, Hamming, and Alignment based Distances

The *edit distance*, also called the Levenshtein distance, between two sequences S_1 and S_2 is defined to be the minimum number of edit operations to transform S_1 to S_2. The edit operations include changing a letter to another, inserting a letter, and deleting a letter. For example, the edit distance between "school" and "spool" is 2 (we need one deletion and one change), and the edit distance between "park" and "make" is 3.

Some authors [126] argue that the edit distance may not be the ideal solution to measure sequence similarity. Consider the following three sequences: aaaabbb, bbbaaaa, and abcdefg. The edit distance between aaaabbb and bbbaaaa is 6, and the edit distance between aaaabbb and abcdefg is also 6. Intuitively aaaabbb and bbbaaaa are more similar to each other than aaaabbb and abcdefg are, but the edit distance cannot tell the difference.

The *Hamming distance* between two sequences is limited to cases when the two sequences have identical lengths, and is defined to be the number of positions where the two sequences are different. For example, the Hamming distance is 1 for "park" and "mark", it is 3 for "park" and "make", and it is 4 between "abcd" and "bcde".

The edit and Hamming distances charge each mismatch one unit of cost in dissimilarity. In some applications where different mismatches are viewed differently, different costs should be charged to different mismatches. Those costs are usually given as a matrix; examples include PAM [122] (a transition probability matrix) and BLOSUM [49] (blocks substitution matrix). Moreover, insertions/deletions can also be charged differently from changes. It is also a common practice to charge more on the first insertion than subsequent insertions, and to charge more on the first deletion than subsequent deletions.

Many fast software tools exist for finding similar sequences through sequence alignment; examples include BLAST and variants [4], FASTA [85], Smith-Waterman [98]. These algorithms compare a given sequences against sequences from a given set, to find homologous (highly similar) sequences. PSI-BLAST [4] and HMM-based methods [59] can be used to find remote homologous (less similar) sequences.

3.3.3 Conditional Probability Distribution based Distance

Reference [126] considers a *conditional probability distribution* (CPD) based distance. The idea is to use the CPD of the next symbol (right after a segment of some fixed length L) to characterize the structural properties of a given sequence (or set of sequences). The distance between two sequences is then defined in terms of the difference between the two CPDs of the two sequences. The similarity between two CPDs can be measured by the *variational distance* or the *Kullback-Leibler divergence* between the CPDs. Let $\mathcal{A}$ be a given alphabet. Let S_1 and S_2 be two sequences. Let Ω be the set of sequences of length L which occur in either S_1 or S_2. Let P_i denote the CPD for S_i. The variational distance between S_1 and S_2 is then defined as

$$\sum_{X \in \Omega, x \in \mathcal{A}} |P_1(x|X) - P_2(x|X)|$$

and the Kullback-Leibler divergence between S_1 and S_2 is defined as

$$\sum_{X \in \Omega, x \in \mathcal{A}} (P_1(x|X) - P_2(x|X)) \cdot log \frac{P_1(x|X)}{P_2(x|X)}.$$

This distance can be easily extended to distance between sets (clusters) of sequences. The computation of CPD based distance can be expensive for large L. The clustering algorithm given in [126] computes the distance of a sequence S with each cluster C; that distance is simulated as the the error of predicting the next symbol of the sequence S using the CPD of the cluster C. Only frequent sequences in a cluster C are used in defining CPDs; if a sequence of length L is not frequent in C, its longest frequent suffix is used in its place.

3.3.4 An Example of Feature based Distance: d2

An example of feature based sequence distance is *d2* [108], which uses k-grams as features. It uses two parameters: a window length W and a gram (word) length w. It considers frequency of all w-grams occurring in a window of length W. The distance between two windows is the Euclidean distance between the two vectors of frequencies of w-grams in the two windows. The $d2$ distance between two sequences is the minimum of the window distances between any two windows, one from each sequence.

GACTTCTATGTCACCTCAGAGGTAGATAGA
CGAATTCTCTGTAACACTAAGCTCTCTTCC

Fig. 3.2. Two DNA sequences

For example, consider the two sequences given in Figure 3.2 for $W = 8$ and $w = 2$. To compute the d2 distance between the two sequences, we need to compute the window distance between all pairs of windows W_1 and W_2. Consider the following pair of windows: $W_1 = ACTTCTAT$ (starting from the second position of the first sequence) and $W_2 = AATTCTCT$ (starting from the third position of the second sequence). The two vectors of frequencies of 2-grams are:

	AA	AC	AG	AT	CA	CC	CG	CT	GA	GC	GG	GT	TA	TC	TG	TT
W_1	0	1	0	1	0	0	0	2	0	0	0	0	1	1	0	1
W_2	1	0	0	1	0	0	0	2	0	0	0	0	0	2	0	1

So $d2(W_1, W_2) = \sqrt{(0-1)^2 + (1-0)^2 + (1-0)^2 + (1-2)^2} = 2$.

We should note that this is a simplified definition; the more general definition allows one to consider an interval of gram lengths.

The d2 distance is not really a distance metric, since it violates the triangle inequality. Consider the three sequences of aabb, abbc, and bbcc for $W = 3$ and $w = 2$. Then $d2(aabb, abbc) = 0$, $d2(abbc, bbcc) = 0$ and $d2(aabb, bbcc) = \sqrt{2}$; so $d2(aabb, abbc) + d2(abbc, bbcc) < d2(aabb, bbcc)$.

3.3.5 Web Session Similarity

Web session sequences are sequences of URLs. Since each URL (see Figure 3.3) can be viewed as a sequence, we need to consider web session sequences as nested sequences. So distance functions over web session sequences (which are sequences of URLs) need to consider similarity between individual URLs (representing web pages) and similarity between sequences of URLs. Reference [118] considers a distance defined following such an approach.

On URL similarity, [118] determines a similarity weight between two given URLs in two steps. In step 1, the two URLs are compared to get the longest common prefix. Here, each level in URLs (a level is a nonempty string between "/") is viewed as a token. In step 2, let M be the length of the longer URL between the two, where each level contributes 1 to the length. Then a weight is given to each level of this longer URL: the last level is given the weight of 1, the second to the last is given the weight of 2, etc. The similarity between the two URLs is defined as the sum of the weight of the longest common prefix divided by the sum of the total weights. Observe that, if the two pages are totally different, i.e. the common prefix is the empty sequence, their similarity is 0.0. If the two URLs are exactly the same, their similarity would be 1.0.

For example, consider the three URLs given in Figure 3.3. URL#1 and URL#2 are more similar to each other than URL#1 and URL#3 are. Similarity between URL#1 and URL#2 is obvious: They are very similar pages with a similar topic about the research work in the Data Mining Laboratory of Wright State University. The similarity between URL#1 and URL#3 is weaker, simply reflecting the fact that both pages come from the same server.

Using the approach discussed above on URL#1 and URL#2, the weights for the tokens of `current.html`, `datamine`, `labs`, and `www.cs.wright.edu` are respectively 1, 2, 3, and 4, and the similarity between URL#1 and URL#2 is $\frac{2+3+4}{1+2+3+4} = \frac{9}{10}$. (The two URLs have equal length. So we can break the tie by saying that URL#1 is the longer one.) For URL#1 and URL#3, the longer URL is still URL#1 and so the weights for the tokens are as above, and the similarity between the two URLs is $\frac{4}{1+2+3+4} = \frac{4}{10}$.

```
URL#1: www.cs.wright.edu/labs/datamine/current.html
URL#2: www.cs.wright.edu/labs/datamine/publications.html
URL#3: www.cs.wright.edu/theses/
```

Fig. 3.3. Three URLs

On URL sequence similarity, URL sequence alignment can be performed through dynamic programming, in order to find the best match between two given URL sequences. The score of an alignment is the sum of the individual weights of the web pages, plus insertion/deletion penalties.

3.4 Classification of Sequence Data

In this section we will discuss several popular classification methods, and their applications to sequence classification problems. We will also briefly discuss several metrics for classification evaluation.

3.4.1 Support Vector Machines

Support vector machines (SVMs) have been one of the most widely used classifiers for sequence classification in recent years. In this section we present the basics of SVM, followed by discussion on how SVM has been used for sequence classification. More details on SVM can be found in [113, 12].

To use SVM for sequence classification, feature vectors are constructed from sequences as input to SVMs. Classification for more than two classes can be reduced to classification for two classes[2], so we discuss SVMs for two-class classification below.

Roughly speaking, SVMs map input vectors to new vectors in a higher dimensional space, where an optimal separating hyperplane (see Figure 3.4) is sought. Points lying on one side of the separating hyperplane belong to

[2] When more than two classes are present, a number of SVMs are built for the application. Two approaches can be used: In the "one class against another class" approach, for each pair of classes we build one SVM to discriminate the two classes. In the "one class against all other classes" approach, for each class we build one SVM to discriminate the class and the union of all other classes. Some method is needed to determine the winning class from the classification results of these classifiers.

one of the two classes, and points lying on the other side belong to the other class. Each side has an associated hyperplane parallel to the separating hyperplane, which contains some training data vectors and which are nearest to the separating hyperplane. The separating hyperplane is chosen to maximize the margin; the margin is defined as the distance between the two parallel planes.

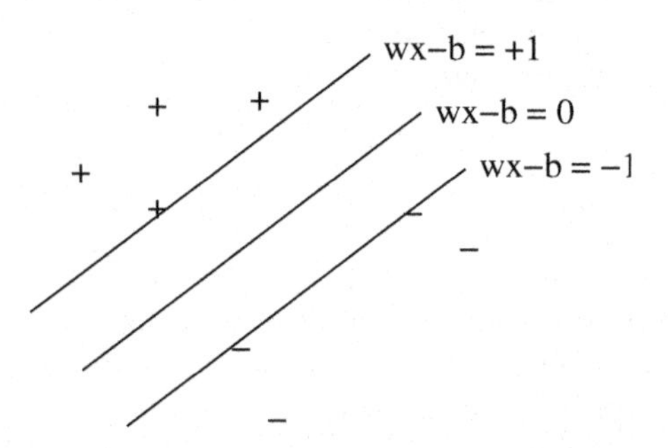

Fig. 3.4. The separating hyperplane and the two support vector hyperplanes of an SVM. Points belong to either the "+" class or the "−" class.

We now briefly discuss the technical details of SVM. Suppose the training data is $\{(\mathbf{x}_1, c_1), ..., (\mathbf{x}_n, c_n)\}$, where $\mathbf{x}_i$ is a vector and c_i is the class of $\mathbf{x}_i$. The two classes are written as -1 and $+1$ to facilitate the expression of optimization objectives in a numerical manner.

We first consider the linear SVM, where we assume that the training data can be separated by a linear hyperplane. Then the separating hyperplane is described by the following equation:

$$\mathbf{w} \cdot \mathbf{x} - b = 0.$$

The vector $\mathbf{w}$ is perpendicular to the separating hyperplane. The data vectors from the two classes nearest to the separating hyperplane are respectively on two parallel hyperplanes, which are described by the following equations:

$$\mathbf{w} \cdot \mathbf{x} - b = +1,$$

$$\mathbf{w} \cdot \mathbf{x} - b = -1.$$

The data points on these two parallel hyperplanes are referred to as the support vectors. The distance between the two hyperplanes is $\frac{2}{|\mathbf{w}|}$; this is the margin which we want to maximize. The optimization can be done through quadratic programming.

When the data vectors cannot be completely separated by a linear hyperplane, we aim to find a separating hyperplane to maximize the margin and minimize the classification error (distance). This two factor optimization problem is converted to a one factor optimization problem through a linear transformation.

The non-linear SVM is similar to the linear SVM, except that each dot product of the form $\mathbf{x} \cdot \mathbf{x}'$ in the quadratic programming conditions is replaced by a kernel function $k(\mathbf{x}, \mathbf{x}')$. The kernel trick has been used often and offers significant computational advantages. Commonly used kernel functions include polynomial ($k(\mathbf{x}, \mathbf{x}') = (\mathbf{x} \cdot \mathbf{x}')^d$ or $k(\mathbf{x}, \mathbf{x}') = (\mathbf{x} \cdot \mathbf{x}' + 1)^d$ for some d), radial basis function ($k(\mathbf{x}, \mathbf{x}') = exp(-\gamma||\mathbf{x} - \mathbf{x}'||^2)$ for some $\gamma > 0$), Gaussian radial basis function ($k(\mathbf{x}, \mathbf{x}') = exp(-\frac{||\mathbf{x}-\mathbf{x}'||^2}{2\sigma^2})$, and sigmoid ($k(\mathbf{x}, \mathbf{x}') = tanh(\kappa\mathbf{x} \cdot \mathbf{x}' + c)$ for some $\kappa > 0$ and $c < 0$).

Applications of SVM for Sequence Classification

The subcellular location of a protein is closely correlated to the function of the protein. A special location-based type of proteins, namely the outer membrane proteins (or OMPs), are of primary research interest for antibiotic and vaccine drug design purposes, as they may be on the surface of bacteria and so are the most accessible targets to develop new drugs against. Several papers use SVMs to predict the subcellular location of a protein or whether a protein is outer membrane protein, based on the sequence of the protein.

- Reference [83] considered 12 subcellular locations in eukaryotic cells. Several kernel functions (the simple linear kernel, the polynomial kernel, and the RBF kernel) were used to build the SVMs, using five different types of features. A SVM is built in one of two manners, either the one location against another location manner, or the one location against all other locations manner. Several types of features were used, namely amino acids, amino acid pairs, one-gapped amino acid pairs (allowing exactly one gap between two amino acids), two-gapped amino acid pairs, and three-gapped amino acid pairs. A data set of proteins with known locations was extracted from the SWISS-PROT database for the experiments.
  Reference [51] used SVM to predict whether a protein is outer membrane, using gapped amino acid pairs as features.
- Reference [95] also used SVM to predict whether a protein is outer membrane, using frequent subsequences as features. It also used the simple linear kernel, the polynomial kernel, and the RBF kernel.
- Reference [100] used multi-kernel SVM for the problem of splice site identification. The main motivation is to produce interpretable SVM classifiers. It achieves this goal by using many kernels, one for each position/length pair. The SVMs can be used to identify the important kernels (e.g. position/length pairs).

3.4.2 Artificial Neural Networks

We first provide some basic information on artificial neural networks (ANN). ANNs are a classification method motivated by biological neural networks. An

ANN can be viewed as a system of a large number of interconnected simple components (the neurons/nodes). A neuron takes the output of some other neurons or some original input to the network as input, and produces its own output. Artificial neural networks can be divided into two major categories, namely the feedforward networks and the feedback/recurrent networks. In the feedforward category the neurons can be divided into layers and a neuron of a given layer only takes the output of neurons at the previous layer as input. In the feedback/recurrent category the networks can be viewed as an arbitrary directed graph with loops. See reference [54] (a tutorial on ANN) for more details.

We now discuss some studies that use ANN to classify sequence data. Reference [123] used ANN for protein sequence family classification. Sequences are mapped to vectors of k-gram frequencies and the vectors are used as input to the ANN. Two methods were used to reduce the dimensionality of k-gram frequency vectors: One reduces the number of letters by mapping amino acids into equivalence groups (as discussed early), and the other uses the SVD method to reduce the dimensionality.

Reference [72] used neural networks and expectation maximization to classify E. coli promoters. Given a sequence, an expectation maximization algorithm is used to locate the -35 and -10 binding sites. Then the nucleotides of 17 positions around the -35 binding site, 11 positions around the -10 binding site, and 7 positions around the transcription start site are used as feature vector for the ANN. In the feature vector, each nucleotide is represented as 4 binary bits.

3.4.3 Other Methods

Reference [134] used Fisher's discriminant algorithm [80] for protein coding gene recognition (in yeast). It represents DNA sequences using the so-called z-curves (which are essentially the cumulative counts of each letter in $\{A, C, G, T\}$ for all sequence positions).

Reference [35] proposed the so-called "segment and combine" approach to classify sequences. First, it builds a set of classifiers; each classifier is built from a sample of subsequences randomly extracted from the sequences given in the original training set. Then an ensemble classifier is built from this set of classifiers.

3.4.4 Evaluation of Classifiers and Classification Algorithms

In this section we briefly discuss the evaluation of classifiers, and the evaluation of classification algorithms. There is a distinction between the two, since a classification algorithm can be used to produce many classifiers, by adjusting the parameters, of a style or approach.

The performance of a classifier is usually measured by classification accuracy, precision and recall. These measures are defined based on a matrix as shown in Table 3.1. The table assumes that there are two classes, referred to as "Positive" and "Negative".

Table 3.1. Confusion Matrix in Classification

	Actual Positive	Actual Negative
Classified as Positive	TP (true positive)	FP (false positive)
Classified as Negative	FN (false negative)	TN (true negative)

The measures of accuracy, precision and recall are defined as follows:

- Accuracy is defined by $(TP + TN)/(TP + FP + FN + TN)$. Accuracy measures the percentage of data from both classes which are correctly classified.
  Very often the two classes of data are not balanced. For example the negative class is usually huge compared to the positive class. Moreover, the costs of different types of classification errors are different. For example, it may be a bigger mistake to classify a cancerous tissue as non-cancerous than the other way around. In such cases, a weighted accuracy measure can be defined.
- Precision of the positive class is defined by $TP/(TP + FP)$. This is the percentage of correctly classified Positive instances among the instances which are classified Positive.
- Recall of the positive class is defined by $TP/(TP + FN)$. This is the percentage of correctly classified Positive instances among the instances which are actual Positive.

For example, consider the case where Positive contains 100 instances and Negative contains 1000. Suppose that, for a classifier f, we have $TP = 50$, $FP = 100$, $FN = 50$, and $TN = 900$. Then the accuracy is $(900 + 50)/(50 + 100 + 50 + 900) = 900/1100 \approx 81.8\%$, the precision of the positive class is $50/(50 + 100) = 50/150 \approx 33.3\%$, and the recall of the positive class is $50/(50 + 50) = 50/100 = 50\%$.

Classifier performance is usually carried out by k-fold cross validation: The available data is randomly shuffled and divided into k parts. For each part, a classifier is built by using the other $k - 1$ parts as training data, and then tested by using the given part as testing data. The average performance over the k classifiers is considered to be the performance of the classifier.

It is common to use the area under the ROC (Receiver Operating Characteristics) curve (AUC) to evaluate the quality of a classification methodology. (Accuracy, precision and recall are measures to evaluate the quality of a classifier.) In the ROC curve one plots a curve over the two-dimensional plane with TP as the Y-axis and FP as the X-axis; the curve is drawn by changing some parameters to yield a range of classifiers with varying (TP, FP) pairs. Classification methods with larger area under the curve are considered better.

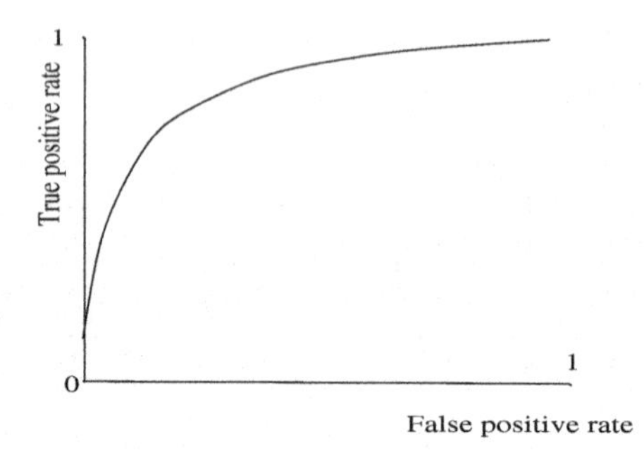

Fig. 3.5. The AUC (area under the ROC curve) as a measure of the quality of a classification algorithm.

3.5 Clustering Sequence Data

Clustering has been widely recognized as a powerful data mining approach. The general goal of clustering is to identify groupings of a given set of objects so that the objects in each group are highly similar to each other and objects in different groups are highly dissimilar to each other. Such groups can capture important phenomenon or concepts for the application under investigation.

In this section we describe several popular sequence clustering algorithms, and briefly discuss the topic of cluster quality evaluation.

3.5.1 Popular Sequence Clustering Approaches

To cluster sequence data, it is necessary to select a suitable clustering algorithm for the data at hand, to select an appropriate distance function, and to address special characteristics of sequence data. Below we will present several popular sequence clustering algorithms, including hierarchical algorithms and graph-based algorithms.[3] These algorithms have been studied for DNA/protein sequence clustering, and for web access sequence clustering.

One important characteristic of sequence data is that sequence similarity can be the result of subsequence similarity in multiple sequence intervals. The implication is that sequence similarity is not transitive, which can lead to unwanted clustering results if not handled properly.

Hierarchical Clustering Algorithms

One of the popular clustering algorithms is the agglomerative hierarchical clustering algorithm [57]. This is a merge-based hierarchical clustering algorithm, which operates in a bottom-up manner. Figure 3.6 gives the pseudo

[3] Besides the two major classes of clustering algorithms discussed here, [99] considers the clustering of sequences using HMM. The paper tries to discover k clusters of sequences, each described by an HMM (see Chapter 4). The whole dataset is then described as a finite mixture of the k HMMs.

code. In contrast, there is also a divisive hierarchical clustering algorithm, which operates in a top-down manner.

Input: a set $\{S_1, ..., S_n\}$ of sequences and a distance function d between sequences;
Output: a hierarchy of clusters;
Method:
1: let $C_i = \{S_i\}$ be a leaf, non-merged cluster for each $1 \leqslant i \leqslant n$;
2: repeat steps (3-5) until just one non-merged cluster remains
3: select a pair C_i and C_j of non-merged clusters such that
$d(C_i, C_j) = min\{d(C_s, C_t) \mid C_s, C_t$ are different non-merged clusters$\}$;
// see below regarding ways to define d on clusters from d on sequences
4: let $C_i \cup C_j$ be a new cluster with C_i and C_j as children;
5: mark C_i and C_j as merged;
6: return the hierarchy of all clusters;

Fig. 3.6. The *Agglomerative Hierarchical Clustering* algorithm.

The hierarchy of clusters is usually shown as a dendrogram (tree). Figure 3.7 gives an example dendrogram.

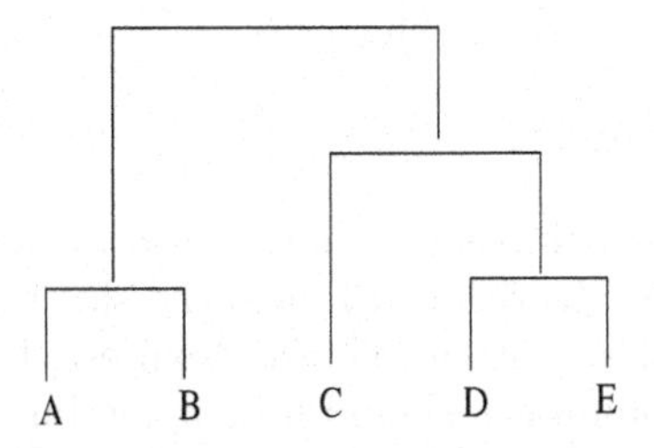

Fig. 3.7. A dendrogram for clustering 5 sequences. There are a total of 9 clusters, including 5 single-sequence clusters. The edges show the parent-child relationship between the clusters. For example, $\{A, B\}$ and $\{D, E\}$ are clusters merged from single-sequence clusters, and $\{C, D, E\}$ is a parent cluster of clusters $\{C\}$ and $\{D, E\}$.

Several issues must be addressed in order to use this algorithm.

(a) How to choose an appropriate distance function.
(b) How to get k clusters given a hierarchy of clusters for a given integer k?
(c) How to define the distance between clusters, given a distance function on individual sequences?

Regarding issue (a), we considered how to define distance functions earlier. The selection decision may be made by considering domain knowledge.

Regarding issue (b), we note that one can get k clusters through a horizontal cut through the hierarchy, removing the part of the hierarchy below the cut. For the hierarchy of clusters in Figure 3.7, one can get 2 clusters from the two children of the root. In general, the sizes of the clusters and their quality are factors that should be considered in determining where the horizontal cut is.

We now turn to issue (c). There are several approaches to defining distance $d(C, C')$ between two clusters C and C' from a distance function d on sequences:

- In the *single-linkage* approach, $d(C, C')$ is defined to be the smallest distance between pairs of sequences, where each pair contains one member of cluster C and one member of cluster C':
$$d(C, C') = min_{S \in C, S' \in C'} d(S, S').$$
- In the *complete-linkage* approach, $d(C, C')$ is defined to be the largest distance between pairs of sequences, where each pair contains one member of cluster C and one member of cluster C':
$$d(C, C') = max_{S \in C, S' \in C'} d(S, S').$$
- In the *average-linkage* approach, $d(C, C')$ is defined to be the average distance between pairs of sequences, where each pair contains one member of cluster C and one member of cluster C':
$$d(C, C') = avg_{S \in C, S' \in C'} d(S, S').$$

The single-linkage approach has been often used for sequence clustering. For example, the d2_cluster algorithm [13] uses the single-linkage Agglomerative Hierarchical Clustering algorithm to cluster expressed sequence tags[4].

The single-linkage approach should be fairly robust in the sense that the final clustering result is not affected by the ordering[5] of the cluster mergers. This property may imply that we can discover arbitrarily shaped clusters: two sequences S and S' will belong to a common cluster if there exists a list of sequences $S = S_1, S_2, ..., S_{m-1}, S_m = S'$ such that $d(S_{i-1}, S_i)$ is relatively small for each i. In contrast, the final result of the complete-linkage approach can be easily affected by the ordering of the mergers.

However, the single-linkage approach has challenges. First, the approach may group the majority of sequences into a single large cluster. A second challenge occurs when the distance function is determined by similarity of subsequence segments and hence sequence similarity is not transitive (see Figure 3.8). This causes problems for the clustering of protein sequences,

[4] An expressed sequence tag or EST is a short sub-sequence of a transcribed spliced nucleotide sequence. ESTs are intended as a way to identify gene transcripts, and are instrumental in gene discovery and gene sequence determination.

[5] Different orders are possible since ties can exist for selecting a pair C_i and C_j of non-merged clusters in step 3 of the hierarchical clustering algorithm.

where the aim is to identify protein families. As illustrated in Figure 3.8, some proteins can belong to multiple domains. Such proteins can induce mergers of two or more clusters (each corresponding to a domain) into one cluster (see Figure 3.9). Reference [30] gives an algorithm to solve this problem, which includes a step to handle multi-domain sequences after single-linkage clustering.

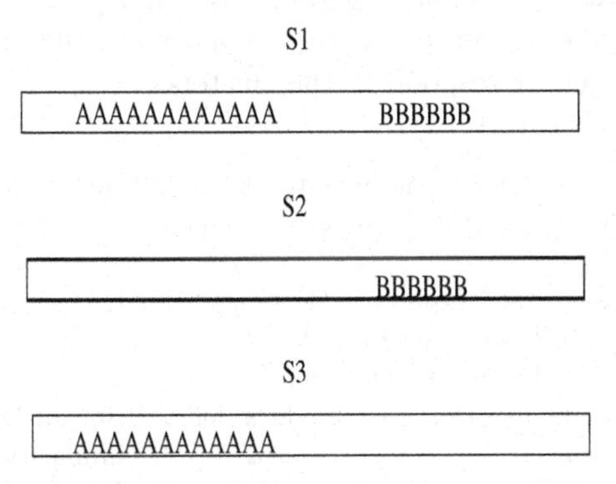

Fig. 3.8. Sequence similarity is not transitive: sequences S_1 and S_2 are similar (sharing a common segment), and sequences S_1 and S_3 are also similar (sharing a common segment), but sequences S_2 and S_3 are not similar.

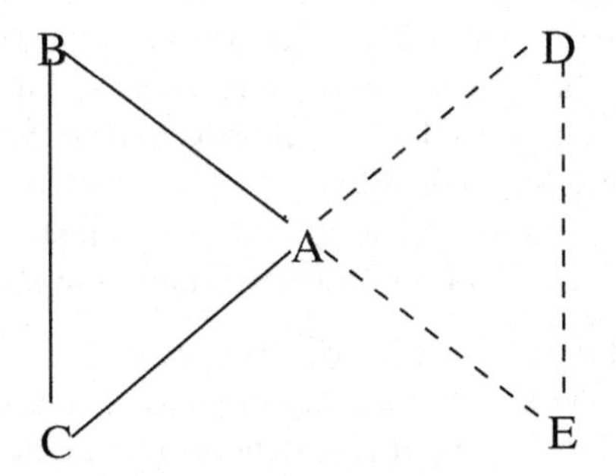

Fig. 3.9. Two domains in one cluster. A,B,C,D, and E are 5 sequences. A,B, and C are similar by sharing a common pattern (a domain motif), and A,D, and E are similar by sharing another common pattern (a domain motif). A is similar to all of the other 4 sequences and hence all 5 sequences are in one cluster under single-linkage clustering. But the five sequences do not share a common pattern.

Graph based Clustering Algorithms

The clustering problem can be naturally cast as a graph optimization problem. Many graph based clustering algorithms have been proposed. In this section we first describe the general framework of graph-based clustering algorithms. Then we discuss several representative variants of such algorithms.

The general framework of graph-based clustering algorithms is given in Figure 3.10. It involves two major parts. In the first part, a weighted graph is constructed from the sequences. In the second part, the graph is segmented into subgraphs which correspond to the clusters.

Input: a set $\{S_1, ..., S_n\}$ of sequences and a distance function d between sequences;
Output: a clustering $C_1, ..., C_m$ of the given sequences;
Method:
1: build an edge-weighted graph $G = (V, E, w)$ from $\{S_1, ..., S_n\}$, where
 - $V = \{S_1, ..., S_n\}$ is the set of nodes,
 - $E \subseteq V \times V$ is the set of edges, and
 - w is the weight function on the edges defined (for all $(u, v) \in E$) by $w(u, v) = max_{(u',v') \in E} d(u', v') - d(u, v)$ (the similarity between u and v);

2: find subgraphs $G_1, ..., G_m$ of G by maximizing total intra-subgraph edge weight (or node similarity) and minimizing total inter-subgraph edge weight;
3: let C_i be the nodes (sequences) in G_i for each i;
4: return the clusters $C_1, ..., C_m$;

Fig. 3.10. The *Graph-based Clustering* algorithm.

For building the weighted graph in graph-based clustering, various distance functions on sequences such as those discussed in Section 3.3 can be used. In addition, sequence alignment algorithms such as BLAST [4] and Smith-Waterman [98] can be used to derive similarity scores. Sometimes, edges with very low similarity weight are ignored.

There are many variants of the graph-based clustering algorithm. They usually differ on the optimization objective that they use.

- Several variants try to minimize the weights of the cross-over edges (u, v) whose end nodes belong to different subgraphs (i.e. u occurs in G_i and v occurs in G_j and $i \neq j$). Let $w(u, v)$ denote the weight of an edge (u, v). For each i, let V_i be the set of nodes of G_i. Let $w(V_i, V_j) = \sum_{u \in V_i, v \in V_j} w(u, v)$, and let $ws = \sum_{1 \leqslant i < j \leqslant m} w(V_i, V_j)$. Observe that $w(V_i, V_j)$ is the sum of weights for edges in $V_i \times V_j$, and ws is the total weights of all edges that connect different subgraphs. The objective is then to find $G_1, ..., G_m$ so that ws is as small as possible.
Reference [60] gives an algorithm for the case $m = 2$, which we call *weighted graph cut minimization* (*WGCM*). *WGCM* starts by randomly dividing V into two sets, V_1 and V_2. Then it performs a number of passes until the

move does not reduce ws. In each pass, it repeatedly checks if ws can be reduced by moving one sequence from V_1 to V_2 or vice versa, and it picks the move that maximizes the reduction. The reduction for moving $v_1 \in V_1$ to V_2 can be computed by

$$w(V_1, V_2) - w(V_1 - \{v_1\}, V_2 \cup \{v_1\}) = \sum_{v_2 \in V_2} w(v_1, v_2) - \sum_{v_1' \in V_1} w(v_1, v_1').$$

(The reduction for moving $v_2 \in V_2$ to V_1 is similar.) In each pass *WGCM* moves each given node at most once, and the number of moves is chosen to maximize the sum of the reductions of the moves.

- Several others try to maximize the density of edges in the subgraphs. Reference [47] uses the notion of edge-connectivity to find highly "coherent" subgraphs. The graph is first converted to a normal graph – only edges whose similarity weights are higher than a threshold are used. The *edge-connectivity* $k(G')$ of a graph G' is the minimum number of edges whose removal results in a disconnected graph. A graph G' is called a *highly connected graph* if $k(G') > \frac{n'}{2}$ edges, where n' is the number of nodes of G'. The paper proposes a so-called *Highly Connected Subgraphs* (*HCS*) algorithm, which produces the clusters by finding the maximal highly connected subgraphs of G.
- Reference [31] considers a Markov Cluster (*MCL*) algorithm to discover good clusters from weighted similarity graphs. The algorithm uses random walk probabilities to guide the search for natural sequence clusters, based on the intuition that random walks should mostly start and end within natural clusters and infrequently lead one from one natural cluster to another natural cluster.

3.5.2 Quality Evaluation of Clustering Results

For clustering, two types of measures of cluster goodness or quality can be used. One type of measures allows us to compare different sets of clusters without reference to external knowledge and is called an internal quality measure. This might be defined as the average pairwise similarity between elements in common clusters. The other type of measures lets us evaluate how well the clustering is working by comparing the groups produced by the clustering techniques to known classes. This type of measure is called an external quality measure. An example of external measures is the F-measure, which combines the precision and recall ideas from information retrieval. The F-measure is basically the harmonic mean of precision and recall:

$$F = \frac{2 * \text{Precision} * \text{Recall}}{\text{Precision} + \text{Recall}}.$$

The last formula is the non-weighted version of F-measure. More details on quality evaluation of clustering can be found in [102].

4

Sequence Motifs: Identifying and Characterizing Sequence Families

This chapter is concerned about sequence motifs. It discusses finding motifs and using motifs in sequence analysis.

A motif is essentially a short distinctive sequence pattern shared by a number of related sequences. The distinctiveness of a motif is mainly reflected in the over representation of the motif pattern at certain locations in the related sequences and the under representation elsewhere (at other locations in the related sequences and at all locations in other sequences). Motifs have been mostly studied in biological sequence analysis applications. In such applications, motifs are believed to be the result of conservation during evolution; the conservation is witnessed by the sharing of the motif by multiple sequences in different individuals, perhaps from different species. Based on that belief, motifs can have important biological functions/structures. This reasoning has been experimentally verified many times.

Roughly speaking, the motif finding task is concerned with site-focused identification and characterization of sequence families. It can be viewed as a hybrid of clustering and classification, and is an iterative process. The identification of sequence families is based on distinctive characteristics of certain short sequence windows; the short windows are identified by local alignment; the local alignment involves some (or all) of the given sequences.

On the other hand, motif analysis is concerned with predicting whether sequences match a certain motif, and the sequence position where the match occurs. Two types of activities can happen here, one involving the scoring of a sequence against a motif, and the other one involving explaining the sequence in terms of the sequence of actions (more formally, states) when the motif "processes" the sequence.

Various motif representations and associated algorithms have been reported. In this chapter we will present some major motif representations. While there have been many algorithms, most of them are instances of one of three algorithmic approaches, namely dynamic programming, expectation maximization, and Gibbs sampling.

Section 4.1 gives the motivations for motif studies, and lists four general motif analysis problems. Section 4.2 discusses major motif representations, together with the simpler algorithmic solutions to the motif finding and motif analysis problems. Section 4.3 presents the non-trivial representative algorithmic approaches to the motif finding and motif analysis problems. Section 4.4 discusses related topics and future research problems.

4.1 Motivations and Problems

In this section we first provide some motivating motif examples from biological applications. Then we define the concept of motif. Finally, we define four motif analysis problems. The solutions to these problems will be considered in the next two sections.

Definition 4.1. *A* motif *is a short distinctive sequence pattern shared by multiple related sequences.* ■

The definition is not very precise. It is only intended to be used as a guideline. The meaning will become clear from the examples given in the next two sections.

4.1.1 Motivations

There are many applications for motifs. Generally speaking, whenever there is a need to identify short distinctive sequence patterns, we have a potential application for motif analysis. Typical applications include the following:

- The identification of boundary/join positions of special types of biological objects represented as sequences. Examples of such objects include genes, introns, exons, promoters, etc. It is of interest to know the starting/ending sites of genes [71], to know the boundary points of introns and exons and points for alternative splicing [97], and to know where the transcription promoters are and where the transcription factor binding sites are [90]. Many more applications exist.
  One possible reason why DNA/RNA motifs are short distinctive sequence patterns is the following: Many biological activities are performed by molecules such as transcription factors and RNA polymerase. The size of such molecules limits the length of the DNA/RNA sequence they can make contact with at once, which may imply that their associated sequence patterns should be short. (It is also conceivable that such molecules cannot have long term memories, another reason for motifs to be short.) Moreover, to conserve energy, they also need to be selective in when to perform biological transformations/actions, which implies that their associated sequence pattern should be distinctive.

- The identification of proteins with certain distinctive topological structures or with certain binding properties or enzymatic activities. The determination of the structure of a protein is a tedious process. One way to get around this difficulty is to use resemblance of a protein with unknown structure to a protein with a known structure, to help determine the structure of the unknown protein [58, 50]. Many important resemblances appear in the form of short sequence patterns. (One possible reason for many patterns that determine protein structures to be short is that an amino acid will only be able to be in contact, either physically or through various attracting/repelling forces, with a limited number of other amino acids.) The use of protein sequence patterns (or motifs) to determine the "active sites" or "functional sites" of proteins is an essential tool of sequence analysis. Databases of such patterns have been created. One such database is PROSITE[1], which consists of documentation entries describing various protein domains, families and functional sites as well as associated patterns and profiles to identify them.

Very often motifs are conserved through the long history of evolution. Many motifs are shared by various species; this fact has been often used in locating motifs by cross-species comparison [78].

4.1.2 Four Motif Analysis Problems

The following are four main problems in the study of motifs:

1. The motif representation problem. The problem is to design various motif representations suitable for different application needs. Depending on the task at hand, different types of motifs with various flexibility might be wanted. For example, it makes biological sense that regulatory sites should allow more freedom, whereas restriction sites should be more rigid [104]. Moreover, an application may have a limitation on computation power. One may want to have a simple motif representation if one has less computation power, and to have a more powerful motif representation when more computation power is available. We will give five types of motif representations below, together with some minor variants.
2. The motif finding problem. In the most general form[2], we are given a set D of sequences, and the goal is to build a motif model for some site shared by some sequences from the given sequences in D. This is a very challenging problem, since one needs to do several things: (a) partition D into two subsets, where D_1 contains the members of the desired family and

[1] http://expasy.org/prosite/

[2] In practice, a user may collect a set of sequences which are believed to contain some important patterns for motif finding. For example, some would pick the upstream promoter regions of a certain length of transcriptional start sites. Such regions may still be very long, and they can contain zero, one or more occurrences of a desired motif.

D_2 contains the non-members of the family; (b) identify a window (the desired site) in each sequence of D_1 (or equivalently, give a local alignment of the sequences in D_1); and (c) build the motif model from the result of (b). The goodness of the motif will be evaluated, and the process may need to be iterated if improvement is desired. There are more possibilities for further complications, depending on, for example, whether a sequence can have at most one occurrence of the motif, or it can have multiple occurrences of the motif.
There are more specific and simpler variants of the problem. For example, a user may indicate that all sequences of D contain matches of the motif, or the user may have provided the local alignment of the sequences, or the user may indicate that each sequence contains at most one occurrence of the desired motif.
Recall that motifs are distinctive sequence patterns. Usually distinctiveness is indicated by having the patterns occur more frequently at the site but infrequently elsewhere. Another approach is to compare the frequency of the pattern at the site against a uniform background distribution which might be estimated from all sequences of interest (e.g. all sequences of the genome of a species). It should be noted that straight frequency counting might be too simple minded; some kind of conditional probability/frequency (relative to another motif/site) is likely needed in practice.
3. Sequence scoring. Given a motif and a sequence, the problem is to compute the score/probability of the sequence being generated by (or matching) the motif. The score can be used to locate new occurrences of a motif. The algorithms for this problem should also be able to identify the likely positions of the occurrences.
4. Sequence explanation. Given a sequence and a motif with hidden states, provide the most likely state path that produced the sequence. While the previous problem is applicable to all motif representations, the current problem is only applicable to motif models with hidden states. The most likely sequence of the hidden states may reveal likely switching points, in the input sequence, from regions of one property (class) to regions of another property (class).

4.2 Motif Representations

In this section we present several major types of motif representations, from simple to complex. Here we also present some simple training algorithms for model building and for sequence scoring, but we leave the more complex algorithms to the next section.

The example given in Figure 4.1, about the -10 region[3] of six E. coli promoters [91], will be used to illustrate many concepts and techniques below.

...TACGAT...
...TATAAT...
...TATAAT...
...GATACT...
...TATGAT...
...TATGTT...

Fig. 4.1. Six Aligned DNA Sequences

4.2.1 Consensus Sequence

For a set of sequences over a sequence alphabet $\mathcal{A}$, a *consensus sequence* is a sequence over sets in $2^{\mathcal{A}} - \{\emptyset\}$. A nonempty subset X of $\mathcal{A}$ is usually written by enclosing the members of X in square brackets. For example, $\{A, C\}$ is written as $[AC]$. The set $X = \mathcal{A}$ can be written as the wild card ".". In addition to sets in $2^{\mathcal{A}} - \{\emptyset\}$, one can also include their complements in order to simplify the consensus sequence. Other shortening notations can also be used, e.g. "$[ST](2)$" is the shorthand for "$[ST][ST]$". If a consensus sequence only uses elements in $\mathcal{A}$ (or singleton sets in $2^{\mathcal{A}}$), then it is rigid in matching; otherwise it is flexible in matching.

For DNA or protein sequences, the IUPAC code (or IUPAC ambiguity code) uses special symbols to represent some important subsets of the DNA/protein alphabet. For example, R represents $[AG]$, S represents $[GC]$, V represents $[ACG]$, etc.

For the sequences in Figure 4.1, TATAAT is a rigid consensus sequence, whereas TAT[AG].T (or TATR.T in IUPAC code) is a flexible one.

Consensus sequences can be viewed, constructed and used as special examples of position weight matrix discussed in the next subsection. So we will leave the discussion on their construction and usage to the next subsection.

4.2.2 Position Weight Matrix (PWM)

Let $\mathcal{A}$ be an alphabet and $w > 0$ be a window width. A *position weight matrix* (PWM), also called *position specific weight matrix* (PSWM) or *position specific scoring matrix* (PSSM), is an $|\mathcal{A}| \times w$ matrix M where $M(x, p)$ is a number, for each $x \in \mathcal{A}$ and $1 \leqslant p \leqslant w$. Each row of M corresponds to a letter in the alphabet $\mathcal{A}$, and each column corresponds to a position in the window of the

---

[3] This region starts at the 10th position to the left of the transcription start site (hence -10); the transcription start site is referred to as position $+1$ [22].

motif. The number $M(x, p)$ is usually proportional to the frequency of x at position p, although it can also be influenced by other factors.

Table 4.1 is an example PWM. Here the values indicate the frequency of letters for given positions.

Table 4.1. A 4×6 PWM for Figure 4.1

A	0	1	0	$\frac{3}{6}$	$\frac{4}{6}$	0
C	0	0	$\frac{1}{6}$	0	$\frac{1}{6}$	0
G	$\frac{1}{6}$	0	0	$\frac{3}{6}$	0	0
T	$\frac{5}{6}$	0	$\frac{5}{6}$	0	$\frac{1}{6}$	1

A consensus sequence can be constructed from a PWM, by including the symbols with high values for each position. For example, the consensus sequence TATAAT can be obtained from the PWM in Table 4.1 by including the most frequent symbol for each position, and the consensus sequence TAT[AG]AT can be obtained by including the symbols whose frequency is above $\geqslant 0.50$ for each position.

Given a PWM M and a sequence S of length w, a *similarity score* $Score(M, S)$ can be defined as $Score(M, S) = \Sigma_{p=1}^{w} M(p, S[p])$. Given a sequence S longer than w and a position $i \leqslant |S| - w + 1$ of S, we can compute a similarity score for the window starting at i. This score can be used to identify positions of sequences as likely occurrences of the motif.

Example 4.2. For the PWM in Table 4.1 and the sequence $S = TATGAT$, we have

$$Score(M, S) = \frac{5}{6} + 1 + \frac{5}{6} + \frac{3}{6} + \frac{4}{6} + 1 = \frac{23}{6}.$$

For sequence $S' = GACAAC$, we have

$$Score(M, S') = \frac{1}{6} + 1 + \frac{1}{6} + \frac{3}{6} + \frac{4}{6} + 0 = \frac{15}{6}.$$

These scores indicate that S is more similar than S' to the motif represented by M. ■

A PWM can also be visualized as sequence logos [94] in a graphical manner. A logo consists of stacks of symbols, one stack for each position in the PWM. The overall height of a stack indicates the sequence conservation at that position, while the height of a symbol indicates the relative frequency of the symbol at that position. In general, a sequence logo can provide a richer and more precise description of, for example, a binding site, than a consensus sequence. Figure 4.2 shows the sequence logo for the PWM in Table 4.1.

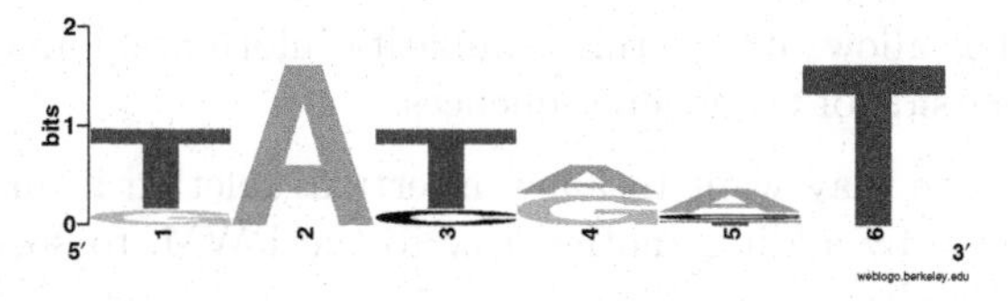

Fig. 4.2. A sequence logo representation of the PWM in Table 4.1.

We now turn to methods for computing a PWM from a given set of aligned sequences. The case for unaligned sequences will be dealt with in Section 4.3.2 using the Gibbs sampling algorithm.

Let $\mathcal{A}$ be an alphabet and w be a window size, and $D = \{S_1, ..., S_N\}$ be a set of N aligned sequences over $\mathcal{A}$. Let a_i be the first position of S_i in the aligned window. Let $n(x,p)$ denote the number of occurrences of x at position p of the window; i.e. $n(x,p)$ is the number of x's among $S_1(a_1+p)$, ..., $S_N(a_N+p)$. The following are three possible methods for computing a PWM. The third one allows us to include similarity between pairs of letters in the definition, while the first two do not.

1. The direct frequency method: In this method, $M(x,p)$ is defined to be the relative frequency of x at position p of the window:
$$M(x,p) = \frac{n(x,p)}{N}.$$
Table 4.1 is computed using this method from the six aligned DNA sequences in Figure 4.1. In the $M(x,p)$ matrix computed by this method, each column is a probability distribution. Moreover the letters with the highest value from the columns form the consensus sequence of the motif.
2. The log odds method: This method defines $M(x,p)$ by
$$M(x,p) = log(Pr(x,p)/b_x) = log(\frac{n(x,p)}{N}/b_x)$$
where b_x is the background frequency of x.
3. The frequency+similarity method: This method defines a PWM by combining a frequency based weighting matrix W and a pairwise similarity matrix Y between the letters in the alphabet. The weighting matrix W can be the PWM defined by either of the two methods above. (It can also include other information. For example, when several of the sequences are very similar, one may want to decrease their importance by giving them less weight.) An example of the matrix Y is Dayhoff's mutational distance matrix [23] between amino acids. For each letter $x \in \mathcal{A}$ and position p, $M(x,p)$ is defined by
$$M(x,p) = \sum_{y \in \mathcal{A}} Y(x,y) \times W(y,p).$$

The use of Y allows us to utilize similarity information among the letters obtained outside of the given sequences.

Sometimes one may want to allow insertion/deletion in motif matching. This can be done by adding another row to the PWM, to specify the insertion/deletion penalty.

In the approaches discussed above, we assumed a zero probability for letters not occurring in a position of the aligned sequences. This might not be very desirable, especially when the training data is small or the the training data collection process is biased. To solve the problem, it is common to use pseudocounts (or Dirichlet priors). For example, the frequency method can be changed to the following: For each position p and each letter y, let $PC(y, p) \geqslant 0$ be a number. Define $M(x, p)$ as follows:

$$M(x,p) = \frac{n(x,p) + PC(x,p)}{N + \sum_{y \in \mathcal{A}} PC(y,p)}.$$

A simple way to define PC is to have $PC(y, p) = 1$ for all y and p. Another way is to have $PC(y, p) = b_y * N$, where b_y is the (background) probability of y occurring in other regions of the sequences or other sequences in the application and N is the number of sequences in D.

Example 4.3. Table 4.2 is the PWM computed using the direct frequency method and the simple pseudocount of 1, from the six aligned DNA sequences in Figure 4.1. Observe that each column is still a probability distribution. ■

Table 4.2. A 4×6 PWM for Figure 4.1 Using Pseudocounts

A	$\frac{1}{10}$	$\frac{7}{10}$	$\frac{1}{10}$	$\frac{4}{10}$	$\frac{5}{10}$	$\frac{1}{10}$
C	$\frac{1}{10}$	$\frac{1}{10}$	$\frac{2}{10}$	$\frac{1}{10}$	$\frac{2}{10}$	$\frac{1}{10}$
G	$\frac{2}{10}$	$\frac{1}{10}$	$\frac{1}{10}$	$\frac{4}{10}$	$\frac{1}{10}$	$\frac{1}{10}$
T	$\frac{6}{10}$	$\frac{1}{10}$	$\frac{6}{10}$	$\frac{1}{10}$	$\frac{2}{10}$	$\frac{7}{10}$

4.2.3 Markov Chain Model

The motif models presented so far have no memory in the sense that the positions in the motif are completely independent of each other. To capture the dependence between positions, we need to add states. Several approaches are possible. This section presents the Markov chain model, the simplest extension to PWM.

In general, for a given data distribution and a given sequence $S = s_1 ... s_m$, one can estimate the probability of S occurring in the distribution using the following:

$$Pr(S) = Pr(s_m \mid s_{m-1}..s_1)Pr(s_{m-1} \mid s_{m-2}..s_1)...Pr(s_2 \mid s_1)Pr(s_1).$$

The first order Markov property assumes that the next state only depends on the present state but not the past states. With that assumption, we can rewrite $Pr(S)$ as follows:

$$\begin{aligned} Pr(S) &= Pr(s_m \mid s_{m-1})Pr(s_{m-1} \mid s_{m-2})...Pr(s_2 \mid s_1)Pr(s_1) \\ &= Pr(s_1)\Pi_{i=2}^{m}Pr(s_i \mid s_{i-1}). \end{aligned}$$

Higher order Markov property assumes that the next state depends on the present state plus some fixed number ($k-1$, where k is the order of the Markov chain) of past states. Below we focus on the first-order Markov model.

A (first-order) *Markov chain model* is a 5-tuple $(\mathcal{A}, Q, q_0, nxt, emt)$ where

- $\mathcal{A}$ is a finite set (the alphabet) of symbols (or letters);
- Q is a finite set of *states*;
- q_0 is the *start* state;
- nxt is the *transition mapping*; it is a total function, assigning probabilities to state pairs in $Q \times Q$, such that the following two conditions are satisfied:
  (a) $0 \leqslant nxt(q, q') \leqslant 1$ for all $q, q' \in Q$, and
  (b) the set of transitions from any given state q defines a probability distribution over the possible next states, i.e.

$$\sum_{q' \in Q} nxt(q, q') = 1;$$

- emt is an *emission function*; it is a partial function from Q to $\mathcal{A}$; q is a *silent* state if $emt(q)$ is not defined, and q is an emitting state otherwise.

For DNA/protein sequences, it is often the case that all non-start states of the Markov models are precisely the letters in the alphabet. Hence the emission function is essentially the identity function.

Example 4.4. We now give an example Markov chain model for the data in Figure 4.1. The alphabet is $\{A, C, G, T\}$. The states are $\{q_0, A, C, G, T\}$, with q_0 being the start state. The emission function is $emt(X) = X$ for $X \in \{A, C, G, T\}$. q_0 is a silent state. The transitions are given in Table 4.3. For brevity, all transitions with probability *zero* are omitted. The transitions can also be given as a $|Q| \times |Q|$ matrix.

■

First-order Markov chain models can be displayed as directed graphs, where the edges have the transition probabilities as labels.

Given a sequence $S = s_1 s_2 ... s_m$ and a Markov chain model M, the *score* of S against M is defined as

$$Score(S, M) = nxt(q_0, s_1)\Pi_{i=1}^{m-1} nxt(s_i, s_{i+1}).$$

Table 4.3. The Transitions in nxt

q_0	A	$\frac{13}{36}$
q_0	C	$\frac{2}{36}$
q_0	G	$\frac{4}{36}$
q_0	T	$\frac{17}{36}$
A	A	$\frac{2}{13}$
A	C	$\frac{2}{13}$
A	T	$\frac{9}{13}$

C	G	$\frac{1}{2}$
C	T	$\frac{1}{2}$
G	A	$\frac{3}{4}$
G	T	$\frac{1}{4}$
T	A	$\frac{8}{11}$
T	G	$\frac{2}{11}$
T	T	$\frac{1}{11}$

Example 4.5. For the Markov model in Table 4.3 and the sequence $S = TATGAT$, we have

$$Score(S, M) = \frac{17}{36} * \frac{8}{11} * \frac{9}{13} * \frac{2}{11} * \frac{3}{4} * \frac{9}{13}.$$

For sequence $S' = GACAAC$, we have

$$Score(S', M) = \frac{4}{36} * \frac{3}{4} * \frac{2}{13} * 0 * \frac{2}{13} * \frac{2}{13}.$$

■

Pseudocounts can also be used to deal with letters with zero observed frequencies, similarly to the case for PWM as discussed in the previous section.

To use Markov chain models to classify two classes, C_1 and C_2, of sequences, we would build two Markov chain models, one for each class. Given a sequence S, we could use the scores of S against the two models to decide the membership of S. Normally, the larger score should indicate the correct class of S.

One might wonder why positions in the motif window are not part of the states in the Markov chain models. In the next section we will see that positions can help, and they have been used in hidden Markov models (discussed in Section 4.2.4).

For Markov chain models, there is a one-to-one correspondence between the emitted (or observed) sequence and the sequence of states. This will no longer be true for the hidden Markov models.

Interpolated Markov (IMM) Motif Models

In a kth-order Markov chain model, each future state depends on the current state and the $k-1$ previous states. The training of such a model may require a large training dataset, even for a moderate value of k. Since it is expensive to collect a large amount of data, it is desirable to make the Markov model more flexible so that less training data is required. Instead of having each future state depending on the current state and the $k-1$ previous states, one can let some future states depend on fewer states when there is not enough data or when fewer states are enough to lead to good performance, and let other future states still depend on k previous states. This is the idea of *interpolated Markov models* [24].

4.2.4 Hidden Markov Model (HMM)

In regular Markov chain models discussed earlier, the states are directly visible to the observer and there is a one-to-one correspondence between the sequence of states and the sequence of emitted symbols. In a hidden Markov model (HMM), these are no longer true: one only observes the emitted symbols but can not observe the sequence of states. These differences arise because a hidden Markov model can emit, in a given state, many symbols in a probabilistic manner, instead of emitting at most one symbol as is the case for a Markov chain model.

Hidden Markov models are useful for situations where certain sequences can be generated from different (biologically) important states/classes. For example, one may want to have one model to describe the probabilities of symbol generation in two classes, where one class is near a site of interest (e.g. gene start site) and another class is far away from that site. In contrast, using Markov chain models one would need to build two or more models, one for each class. By finding the most likely state sequence for a symbol sequence in a given HMM, one can identify the points where biological important switches occurred.

A *hidden Markov model* is a 6-tuple $(\mathcal{A}, Q, q_0, q_e, nxt, emt)$ where

- $\mathcal{A}$ is a finite set (the alphabet) of symbols (or letters);
- Q is a finite set of *states*;
- q_0 is the *start* state and q_e is the *end* state;
- nxt is the *transition mapping*, which is a total function assigning probabilities to state pairs in $Q \times Q$, satisfying the following two conditions:
  (a) $0 \leqslant nxt(q, q') \leqslant 1$ for all $q, q' \in Q$, and
  (b) the set of transitions from any given state q defines a probability distribution over the possible next states, i.e.

$$\sum_{q' \in Q} nxt(q, q') = 1;$$

- emt is the *emission mapping*, which is a partial function assigning probabilities to pairs in $Q \times \mathcal{A}$, satisfying the following two conditions:
  (a) $0 \leqslant emt(q, x) \leqslant 1$ if it is defined, for all $q \in Q$ and $x \in \mathcal{A}$,
  (b) for any state q such that $emt(q, y)$ is defined for some $y \in \mathcal{A}$, it is the case that $emt(q, x)$ is defined for all $x \in \mathcal{A}$ and that

$$\sum_{x \in \mathcal{A}} emt(q, x) = 1.$$

A state q is a *silent* state if $emt(q, x)$ is not defined for any $x \in \mathcal{A}$, and q is an *emitting* state otherwise.

Example 4.6. A simple example is the following HMM $M_{rainsun}$. For brevity, we omit the transition/emission entries with zero probabilities.

- The alphabet consists of three observations: *walk*, *shop*, *clean*.
- The states are: *start*, *rainy*, *sunny*, *end*.
- The transition probabilities are:

$$\begin{aligned}
&nxt(start, rainy) = 0.6,\\
&nxt(start, sunny) = 0.4,\\
&nxt(rainy, rainy) = 0.6,\\
&nxt(rainy, sunny) = 0.3,\\
&nxt(rainy, end) = 0.1,\\
&nxt(sunny, rainy) = 0.4,\\
&nxt(sunny, sunny) = 0.5,\\
&nxt(sunny, end) = 0.1,\\
&nxt(end, end) = 1.
\end{aligned}$$

- The emission probabilities are:

$$\begin{aligned}
&emt(rainy, walk) = 0.1,\\
&emt(rainy, shop) = 0.6,\\
&emt(rainy, clean) = 0.3,\\
&emt(sunny, walk) = 0.3,\\
&emt(sunny, shop) = 0.5,\\
&emt(sunny, clean) = 0.2.
\end{aligned}$$

■

Since there is no one-to-one correspondence between the sequence of emitted letters and the sequence of states, we will need to distinguish the two sequences. In fact, two of the three main problems for HMM deal with the determination of the optimal correspondence between the two sequences. The three main problems are:

- Given an HMM model and a sequence, compute the probability that the sequence is emitted by the HMM. This problem is solved by the forward-backward procedure discussed in Section 4.3.1.
- Given an HMM model and a sequence, find the most likely sequence of hidden states that could have emitted the given sequence. This problem is solved by the Viterbi algorithm discussed in Section 4.3.1.
- Given an output sequence or a set of such sequences, find the most likely state transition and output probabilities. In other words, train the parameters of the HMM given a dataset of sequences. This problem is solved by the Baum-Welch algorithm discussed in Section 4.3.3.

Reference [92] is a tutorial on hidden Markov models, containing discussion on related topics and applications of HMM.

The construction of an HMM from training data can be costly, involving both the search of the states, and the parameters for the transitions between the states and the emission of symbols. The topology of an HMM, namely the number of states and how they are linked through transitions, is referred to as the architecture of the HMM. It may help by separating the choice of the architecture and the determination of the parameters. This approach has been taken many times; the next subsection provide more details on this approach.

Profile HMM Model

Special HMM architectures can limit the complexity, and hence speed up the training time, of the HMM. Moreover, a simple structure implies that the associated model has a small number of parameters and can be trained on a small amount of data. Special HMM architectures may also lead to good models if they fit the structure of the application nicely.

An example HMM architecture is *profile HMM* [62], which is well suited as a motif representation for interesting sites in sequences. Profile HMM is also called *linear HMM* [38].

The structure of the profile HMMs is suitable as a motif representation for sites for two reasons: it has a linear structure to match that of a site, and it attempts to reflect the process of evolution. The backbone of a profile HMM is a sequence of states, called "match states," which represent the canonical sequence (rigid consensus sequence) for the family under consideration. Each match state corresponds to one position in the canonical sequence. The series of states is similar to a PWM (also called a profile), since each state contains a frequency distribution across the entire alphabet. The linear topology implies that, once a state has been traversed, it cannot be entered a second time.

To model the process of evolution, two additional types of states, insert and delete states, are included in profile HMM. One delete state lies in parallel with each match state and allows the match state to be skipped. Since delete states do not emit characters, aligning a sequence to a delete state corresponds to the sequence having a deletion at that position. Insert states with self-loops are juxtaposed between match states, allowing one or more bases to be inserted between two match states.

4.3 Representative Algorithms for Motif Problems

In this section we present several main algorithmic techniques for solving several motif finding and motif analysis problems. These include dynamic programming, Gibbs sampling, and expectation maximization. These techniques will be used to solve the problems of sequence scoring given an HMM model, sequence explanation (most likely state path) given an HMM model, motif finding for the PWM representation, and motif finding for the HMM representation.

4.3.1 Dynamic Programming for Sequence Scoring and Explanation with HMM

In this section we solve two problems, sequence scoring and sequence explanation (most likely state path), using the dynamic programming technique. The algorithms are respectively called the "Forward" algorithm and the "Viterbi" algorithm. More detailed discussion on these algorithms can be found in many tutorials on HMM, e.g. [92].

Dynamic programming is a very powerful algorithmic approach. It solves a problem by decomposing the problem into many subproblems, solving the subproblems one by one, and building the solution of the larger problems by using the answers to smaller ones, until all of the problems are solved.

We first consider the sequence scoring problem. Suppose we are given an HMM $M = (\mathcal{A}, Q, q_0, q_e, nxt, emt)$ and a sequence $S = s_1...s_n$. Our goal is to compute the probability of S being emitted by M. In brute force computation, one will enumerate all possible state paths that can generate S, compute the probability of S being generated by each such path, and then sum up those probabilities. The brute force approach is too expensive.

In the dynamic programming approach, we consider the following small subproblems. For each $1 \leqslant i \leqslant n$ and each state $q \in Q$, we will compute the probability of M being in state q immediately after the time when the first i symbols in the sequence (namely the prefix $s_1...s_i$) have been emitted. Let $f_q(i)$ denote that probability. For the original scoring problem, we want to compute $f_{q_e}(n)$. The dynamic programming algorithm is shown in Figure 4.3.

Input: an HMM $M = (\mathcal{A}, Q, q_0, q_e, nxt, emt)$ and a sequence $S = s_1...s_n$;
Output: the probability of S being generated by M;
Method:
1: let $f_{q_0}(0) = 1$, and let $f_q(0) = 0$ for each silent state q;
2: for $i = 1..n$ do
3: let $f_{q'}(i) = emt(q', s_i) \sum_q f_q(i-1)\ nxt(q, q')$ for each emitting state q';
4: let $f_{q'}(i) = \sum_q f_q(i)\ nxt(q, q')$ for each silent state q';
5: return $Pr(S) = f_{q_e}(n)$.

Fig. 4.3. The *Forward* algorithm.

Example 4.7. Let $M_{rainsun}$ be the HMM given in Example 4.6, and let $S =$ *shop, walk*. To find the probability for S, the algorithm will compute many other probabilities. We will only show the computation of the subproblems that contributed to the final answer.

$f_{start}(0) = 1,$
$f_{rainy}(1) = 0.6 * 0.6 = 0.36,$
$f_{sunny}(1) = 0.5 * 0.4 = 0.2,$
$f_{rainy}(2) = 0.1 * (0.36 * 0.6 + 0.2 * 0.4) = 0.0296,$
$f_{sunny}(2) = 0.3 * (0.36 * 0.3 + 0.2 * 0.5) = 0.0624,$
$f_{end}(2) = 0.0296 * 0.1 + 0.0624 * 0.1 = 0.0092.$

So the probability of S being generated by $M_{rainsun}$ is 0.0092.

For any integer $n \geqslant 1$, it can be shown that $\sum_{|S|=n} Pr(S, M) = 1$. Since there are a total of 9 paths of length 2, $S = shop, walk$ can not be the most probable emitted sequence of length 2 for $M_{rainsun}$. ■

Next we consider the sequence explanation problem. Suppose we are given an HMM $M = (\mathcal{A}, Q, q_0, q_e, nxt, emt)$ and a sequence $S = s_1...s_n$. Our goal is to compute the most likely state path that generated S. In the dynamic programming approach, let $mp_q(i)$ denote the probability of the most probable state path that ends in state q after generating exactly the prefix $s_1...s_i$. To answer the original problem, we want to get $mp_{q_e}(n)$ and want to get the most probable path. The dynamic programming algorithm computes the answers to $|Q| \times n$ subproblems. The algorithm is called the "Viterbi" algorithm and its pseudo-code is given in Figure 4.4. The algorithm is very much like the *Forward* algorithm, except that pointers $ptr_q(i)$ are used for retrieving the most probable path at the end of the computation (step 11). It returns the most probable path and its probability.

Input: an HMM $M = (\mathcal{A}, Q, q_0, q_e, nxt, emt)$ and a sequence $S = s_1...s_n$;
Output: the most probable state path that generated S;
Method:
1: let $mp_{q_0}(0) = 1$, and let $mp_q(0) = 0$ for each emitting state q;
2: for $i = 1..n$ do
3: for each emitting state q'
4: let $mp_{q'}(i) = emt(q', s_i) \max_q[mp_q(i-1)\ nxt(q, q')]$;
5: let $ptr_{q'}(i) = \text{argmax}_q[mp_q(i-1)\ nxt(q, q')]$;
6: for each silent state q'
7: let $mp_{q'}(i) = \max_q[mp_q(i)\ nxt(q, q')]$;
8: let $ptr_{q'}(i) = \text{argmax}_q[mp_q(i)\ nxt(q, q')]$;
9: let $Pr(S, \pi) = \max_q[mp_q(n)\ nxt(q, q_e)]$;
10: let $\pi_n = \text{argmax}_q[mp_q(n)\ nxt(q, q_e)]$;
11: for $i = n-1...1$ let $\pi_i = ptr_{\pi_{i+1}}(i)$;
12: return π as the most probable state path and $Pr(S, \pi)$ as its probability.

Fig. 4.4. The *Viterbi* algorithm.

Example 4.8. Let $M_{rainsun}$ be the HMM given in Example 4.6, and let $S = shop, walk$. To find the most probable path for S, the algorithm will compute many probabilities. We will only show the computation of the subproblems that contributed to the final answer.

initialization:
$mp_{start}(0) = 1$, $mp_{rainy}(0) = 0$, $mp_{sunny}(0) = 0$,

$i = 1$, emitting states:
$$mp_{rainy}(1) = emt(rainy, shop) * max_q[mp_q(0) * nxt(q, rainy)]$$
$$= 0.6 * 0.6 = 0.36,$$
$$mp_{sunny}(1) = emt(sunny, shop) * max_q[mp_q(0) * nxt(q, sunny)]$$
$$= 0.5 * 0.4 = 0.20,$$
$$ptr_{rainy}(1) = ptr_{sunny}(1) = start,$$

$i = 1$, silent states:
$$mp_{start}(1) = max_q[mp_q(1) * nxt(q, start)] = 0,$$
$$mp_{end}(1) = max_q[mp_q(1) * nxt(q, end)] = 0.36 * 0.1 = 0.036,$$
$$ptr_{end}(1) = rainy,$$

$i = 2$, emitting states:
$$mp_{rainy}(2) = emt(rainy, walk) * max_q[mp_q(1) * nxt(q, rainy)]$$
$$= 0.1 * max(0.36 * 0.6, 0.20 * 0.4) = 0.0216,$$
$$mp_{sunny}(2) = emt(sunny, walk) * max_q[mp_q(1) * nxt(q, sunny)]$$
$$= 0.3 * max(0.36 * 0.3, 0.2 * 0.5) = 0.0324,$$
$$ptr_{rainy}(2) = ptr_{sunny}(2) = rainy.$$

So the most probable path for generating S is $start, rainy, sunny$, and the probability of the path is 0.0324. ■

4.3.2 Gibbs Sampling for Constructing PWM-based Motif

This section has two goals. First we answer the following question:

How to construct a PWM motif from a set of (unaligned) sequences?

Second, we present the Gibbs sampler as a search algorithm for motif finding.

Specifically, we first present the Gibbs-sampling based algorithm for dynamically aligning multiple sequences and finding a PWM motif from them, for a fairly restricted scenario where each sequence in the input contains exactly one occurrence of the motif to be found. Then we discuss extensions for finding motifs from general input sequences without those constraints.

We begin with some brief background information on Gibbs sampling. More detailed discussion can be found in [15]. Gibbs sampling is an algorithm to generate a sequence of samples from the joint probability distribution of $N \geqslant 2$ random variables. It is a special case of the Metropolis-Hastings algorithm, and thus an example of a Markov chain Monte Carlo algorithm. The sampling is performed over one particular random variable at a time, while keeping the other random variables fixed. Sampling in the entire space is achieved by round-robin one-variable sampling (see the next paragraph) until some stopping condition is reached. Given a function f, by approximating the probability of the samples X to be proportional to $f(X)$, we can use Gibbs sampling to search for the point X where $f(X)$ is maximal. The algorithms that we will discuss below are examples of this idea.

To illustrate round-robin one-variable sampling of Gibbs sampling, suppose we have three variables, (x_1, x_2, x_3). Let $Pr(x_i \mid x_j, x_k)$ denote the conditional probabilities. Let $(x_1^{(0)}, x_2^{(0)}, x_3^{(0)})$ be the initial values of the three variables at round zero. The superscript denotes the "round number." For round $t+1$, Gibbs sampler draws[4] $x_1^{(t+1)}$ from the conditional distribution of $Pr(x_1 \mid x_2^{(t)}, x_3^{(t)})$, draws $x_2^{(t+1)}$ from $Pr(x_2 \mid x_1^{(t+1)}, x_3^{(t)})$, and draws $x_3^{(t+1)}$ from $Pr(x_3 \mid x_1^{(t+1)}, x_2^{(t+1)})$.

The first algorithm we discuss is from [63]. The input to the algorithm is a set of N sequences $S_1, ..., S_N$ over an alphabet $\mathcal{A}$, and a fixed window width W. The algorithm assumes that each sequence contains exactly one occurrence of the motif to be found, and the motif contains no gap. The algorithm maintains three evolving data structures. The first is the PWM $Q(i, j)$, a $|\mathcal{A}| \times W$ matrix describing the motif to be found. The second is a probabilistic description $P(x)$ about the background frequencies of symbols of $\mathcal{A}$ in regions not described by the motif. The third data structure is an array $Posit(i)$ that describes the sites of the motif in the sequences: $Posit(i)$ is a starting position of the motif in sequence S_i.

The objective is to identify the most probable common pattern among the N sequences. This pattern is obtained by locating the alignment that maximizes the ratio of the probability of the pattern to background probability. The algorithmic process to achieve this is through many iterations of two steps of the Gibbs sampler.

In this Gibbs-sampling based algorithm, we treat each sequence S_i as a random variable, and we treat each position in the sequence as a possible sample. A sample is weighted by the ratio of its score in the motif Q to the ratio of its score in the background model P. Since Gibbs sampling maximizes (locally at least) the ratio, the final motif found should be fairly common in the selected sites of the given sequences but not in other windows of the sequences.

The algorithm can be stopped after a large number of iterations (e.g. 500 or 1000). Since the computation is non-deterministic, it is advisable to run the algorithm multiple times.

Several extensions to the above algorithm have been reported, including some reported in [63]. In particular, reference [69] considers the following extensions:

---

[4] When drawing samples from a distribution, the probability of a particular value v being drawn is proportional to the probability of v in the given distribution.

Input: a set of N sequences $S_1, ..., S_N$ over $\mathcal{A}$, and a window width W;
Assumptions: each sequence contains one occurrence of the motif to be found;
pseudocounts b_x ($x \in \mathcal{A}$) are also given (with 1 as default values);
Output: a PWM describing a motif for the N sequences;
Method:
1: Initialize *Posit* by choosing a random starting position for each sequence S_i;
2: Repeat the Predictive Update and Sampling steps:
Predictive Update Step:
3: one of the N sequences, S_z, is chosen at random (or in round-robin fashion);
4: update Q and P as follows:
5: let $Q(x, j) = \frac{c_{x,j}+b_x}{N-1+B}$, where c_{xj} ($x \in \mathcal{A}$, $1 \leqslant j \leqslant W$) is the count of x in the multi-set $\{S_i(Posit(i)+j) \mid i \neq z\}$ and $B = \sum_{x \in \mathcal{A}} b_x$;
6: let $P(x)$ be $\frac{r_x}{t}$, where r_x is the number of occurrences of x in non-motif positions (namely positions j of S_i such that $i \neq z$ and $[j < P(i)$ or $j \geqslant P(i) + W])$ of the $N-1$ sequences and t is the total number of non-motif positions;
Sampling Step:
7: for each position k ($1 \leqslant k \leqslant |S_z| - W + 1$) of S_z, let $A(k) = \frac{Score_Q(k)}{Score_P(k)}$, where $Score_Q(k)$ is the probability of generating $S_z(k)...S_z(k+W-1)$ by Q and $Score_P(k)$ is that of generating $S_z(k)...S_z(k+W-1)$ by P;
8: randomly select one k_0 based on the weighting $A(k)$ for all k, and let $Posit(z) = k_0$;

Fig. 4.5. The *Gibbs-Sampling PWM Finding* algorithm.

- It considers the situation where some sequences in the input may have no copies of a motif while others may have multiple copies. This is an important relaxation, since the target family of sequences characterized by the motif will have to be discovered, together with the local alignment and the motif.
- It also studies how to find motifs which are located in two or more blocks (which can be separated by gaps), and to find palindromic motifs.
- To better capture the characteristics of local sequence environment, a high-order Markov model for the background is adopted (which looks at successive duplet or triplet of symbols at a time).

4.3.3 Expectation Maximization for Building HMM

In this section we present the expectation maximization technique, for the problem of "How to build an HMM from a set of unaligned training sequences?" More detailed discussion on this algorithm can be found in many tutorials on HMM, e.g. [92].

The construction of HMM from given sequences is made harder because the states are hidden. The hidden states make the counting of state transitions and of symbol emissions difficult. In the expectation maximization approach, we get these counts by computing their expected values (defined as the probabilistic average of their counts over all possible state paths).

The expectation maximization based algorithm is called the *Baum-Welch* algorithm. It is also called the *Forward-Backward* algorithm, because it combines both the *Forward* algorithm and the *Backward* algorithm (the latter is a dual of the former). The main steps of the algorithm are given in Figure 4.6. We discuss each of these steps below.

Input: a set D of sequences over an alphabet $\mathcal{A}$;
Output: an HMM;
Method:
1: initialize the parameters of the HMM;
2: repeat the following two steps until convergence
3: compute the expected number of occurrences of each state transition and the expected number of occurrences of each symbol emission;
4: adjust the parameters to maximize the likelihood of these expected values;

Fig. 4.6. The *Baum-Welch* algorithm.

The initialization step will need to determine the number of states, and the transition and emission probabilities. One needs to fix the desired number of states. The transition and emission probabilities can be initialized randomly.

In the expectation step (step 3), we need to compute, for each sequence $S = s_1...s_n$ in D, the probability of the ith symbol being produced by state q in some state path: $Pr(q, i \mid S)$. By definition, $Pr(q, i \mid S) = \frac{Pr(q,i,S)}{Pr(S)}$. Now,

$$Pr(q, i, S) = Pr(s_1...s_i, \text{ state } = q)Pr(s_{i+1}...s_n \mid \text{ state } = q),$$

where $Pr(s_1...s_i, \text{ state } = q)$ is the probability that the HMM is at state q after the prefix $s_1...s_i$ has been emitted, and $Pr(s_{i+1}...s_n \mid \text{ state } = q)$ is the probability that the suffix $s_{i+1}...s_n$ is emitted after the HMM is at state q.

Observe that $Pr(s_1...s_i, \text{ state } = q)$ is precisely $f_q(i)$ defined earlier and can be produced by the *Forward* algorithm. We define $b_q(i) = Pr(s_{i+1}...s_n \mid \text{ state } = q)$, which can be computed by the *Backward* algorithm given in Figure 4.7.

Using $f_q(i)$ and $b_q(i)$, we can rewrite $Pr(q, i \mid S)$ as

$$Pr(q, i \mid S) = \frac{Pr(q, i, S)}{Pr(S)} = \frac{f_q(i)b_q(i)}{Pr(S)}.$$

We can now estimate the expected numbers for transitions and emissions. Let $c(q, s)$ denote the expected number of times symbol s is emitted at state q, and let $c(q, q')$ denote the expected number of times the transition from state q to q' occur in "processing" S. Then

Input: an HMM $M = (\mathcal{A}, Q, q_0, q_e, nxt, emt)$ and an emitted sequence $S = s_1...s_n$;
Output: the probabilities $b_q(i)$;
Method:
1: let $b_q(n) = nxt(q, q_e)$ for states q with a transition to the end state q_e;
2: for $i = n..1$ do
3: let $b_{q'}(i) = \sum_{q\ is\ emitting\ state} emt(q, s_{i+1})\ b_q(i+1)\ nxt(q', q)$
$+ \sum_{q\ is\ silent\ state} b_q(i)\ nxt(q', q)$;
4: return[5] the set of probabilities $\{b_q(i) \mid q \in Q, 1 \leqslant i \leqslant n\}$.

Fig. 4.7. The *Backward* algorithm.

$$c(q, s) = \sum_{S \in D} \frac{1}{Pr(S)} \sum_{S[i]=s} f_q(i) b_q(i),$$

$$c(q, q') = \sum_{S \in D} \frac{1}{Pr(S)} \sum_{1 \leqslant i < |S|} f_q(i) nxt(q, q') emt(q', S[i+1]) b_{q'}(i+1)$$

if q' is an emitting state, and

$$c(q, q') = \sum_{S \in D} \frac{1}{Pr(S)} \sum_{1 \leqslant i < |S|} f_q(i) nxt(q, q') b_{q'}(i+1)$$

if q' is a silent state.

In the expectation step (step 4), we use the expected numbers to derive the maximum likelihood estimators:

$$emt(q, s) = \frac{c(q, s)}{\sum_{s'} c(q, s')},$$

$$nxt(q, q') = \frac{c(q, q')}{\sum_{q''} c(q, q'')}.$$

Pseudocounts can also be used to avoid zero probabilities.

The maximum likelihood estimators are parameter values in a model under consideration that optimize the likelihood that the model produced the data. That is, we look for solutions to the following expression: $\text{argmax}_\lambda Pr(D \mid \lambda)$.

The *Baum-Welch* algorithm will find locally optimal parameters. It is not guaranteed to find the globally optimal solution. Luckily, for many situations, the locally optimal solutions can produce fairly satisfactory HMM models.

4.4 Discussion

It is of interest to compare two motifs to see how similar they are. The problem can be cast as the search of the optimal alignment of two motifs. This problem was considered for PWM in [26]. The paper considered various scoring functions using similarity measures between two score vectors over a given alphabet, including dot product, averaging, Kullback-Leibler divergence, symmetrized entropy, Jensen-Shannon divergence, Euclidean distance, Pearson correlation, etc.

We note that the algorithms discussed in this chapter only find one or several motifs of a given model from a given dataset. One research question is the following: Are there efficient algorithms for finding all possible motifs of a given model from a given dataset, for given thresholds?

For general data, there have also been studies similar to the topic of this chapter, where data clustering is combined with the construction of models of the clusters. Examples include conceptual clustering [79, 32] and succinct and discriminative description of clustering [16].

5

Mining Partial Orders from Sequences

A major type of information to be discovered from sequence data mining comes from finding the ordering information hidden in the data. In Chapter 2, we illustrated sequential pattern mining which finds the common subsequences shared by many sequences in question. However, sequential patterns may not be sufficient in disclosing the ordering information in some applications.

Example 5.1. Suppose the students in a part-time technical certificate program need to take the six courses in Table 5.1. A student takes only one course at a time.

Course-id	Course title
PG	Programming
DS	Data Structures
SE	Software Engineering
DB	Database Management Systems
IR	Information Retrieval
DM	Data Mining

Table 5.1. The set of courses in Example 5.1.

Let us analyze the students' sequences of taking the courses in Table 5.2 to investigate whether there are some orders which students follow.

Finding the frequent patterns may help to capture the possible "templates" of students' course taking. Naturally, we may try mining sequential patterns. Let the minimum support threshold $min\_sup = 3$. The following four sequences are sequential patterns since they are subsequences of three sequences, 1, 2 and 4 in the database.

$$\text{PG} \rightarrow \text{DS} \rightarrow \text{DB} \rightarrow \text{IR};$$
$$\text{PG} \rightarrow \text{DS} \rightarrow \text{DB} \rightarrow \text{DM};$$
$$\text{PG} \rightarrow \text{SE} \rightarrow \text{DB} \rightarrow \text{IR};$$

Student-id	Course sequence
1	PG → DS → SE → DB → IR → DM
2	PG → SE → DS → DB → IR → DM
3	DS → PG → DM → IR → SE
4	PG → DS → SE → DB → DM → IR

Table 5.2. The sequences of taking courses.

$$\text{PG} \rightarrow \text{SE} \rightarrow \text{DB} \rightarrow \text{DM}$$

However, the sequential patterns cannot completely capture the ordering shared by students 1, 2 and 4. It is easy to see that the ordering shared by these three students is the partial order R shown in Figure 5.1.

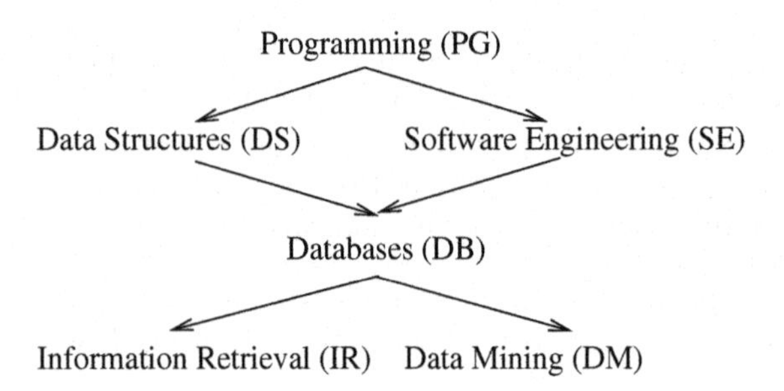

Fig. 5.1. A frequent partial order R in Example 5.1.

We can make two interesting observations. First, the partial order R summarizes the four sequential patterns – the four sequential patterns are paths in R. Second, the partial order R provides more information about the ordering than the sequential patterns. For example, R indicates that some students often take Software Engineering and Data Structures in any order, but those two courses are taken before Databases. This may reflect that Data Structures and Software Engineering may be the prerequisites of Databases. Such information is not presented in the sequential patterns explicitly.

Partial order R is shared by a good number of students and is meaningful in the application. For example, after a student finishes the Programming course, a program advisor may provide the students more information on Data Structures and Software Engineering. However, at this point, the information about Databases is not needed yet.

This example motivates the idea of using frequent partial orders to effectively summarize sequential patterns and provide more general and more concise ordering information. ■

In this chapter, we discuss how to discover partial orders from sequence data. We introduce two types of methods. First, we discuss a method to mine frequent closed partial orders from strings. Second, we discuss how to find the best partial order that is shared by the majority in a set of sequences.

5.1 Mining Frequent Closed Partial Orders

Let us first look at the problem of finding frequent partial orders in a large set of strings.

5.1.1 Problem Definition

A *partial order* is a binary relation that is reflexive, antisymmetric, and transitive. A *total order* (also called linear order) is a partial order R such that for any two items x and y, if $x \neq y$ then either $R(x, y)$ or $R(y, x)$ holds.

A partial order R can be expressed as a directed acyclic graph (DAG for short): the items are the vertices in the graph and $x \rightarrow y$ is an edge if and only if $(x, y) \in R$ and $x \neq y$. We also write an edge $x \rightarrow y$ as (x, y) or xy. For example, Figure 5.2(a) shows a partial order R, which has 13 edges.

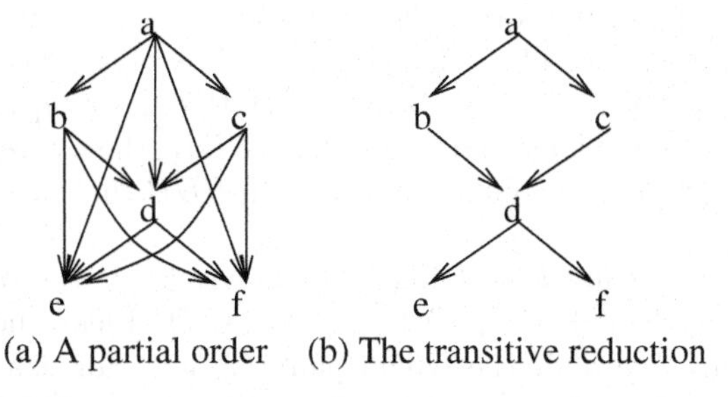

(a) A partial order (b) The transitive reduction

Fig. 5.2. A partial order and its transitive reduction.

Since a partial order is transitive, some edges can be derived from the others and thus are redundant. For example, in Figure 5.2(a), edge $a \rightarrow d$ is redundant given edges $a \rightarrow b$ and $b \rightarrow d$. Generally, an edge $x \rightarrow y$ is *redundant* if there is a path from x to y that does not contain the edge. For a partial order R, the *transitive reduction* of R can be drawn in a *Hasse diagram*: for $(x, y) \in R$ and $x \neq y$, x is positioned higher than y; edge $x \rightarrow y$ is drawn if and only if the edge is not redundant. Figure 5.2(b) shows the transitive reduction of the same partial order R in Figure 5.2(a). The transitive reduction has only 6 edges. For an order R, the transitive reduction may have much fewer edges.

In this chapter, we draw a partial order in a Hasse diagram, that is, its transitive reduction, and omit the isolated vertices. For example, Figure 5.3 shows four partial orders R_1, R_2, R_3 and R_4, and R_1 is further a total order.

Let V be a set of items, which serves as the domain of our string database. A *string* defines a global order on a subset of V. Here, we focus on strings

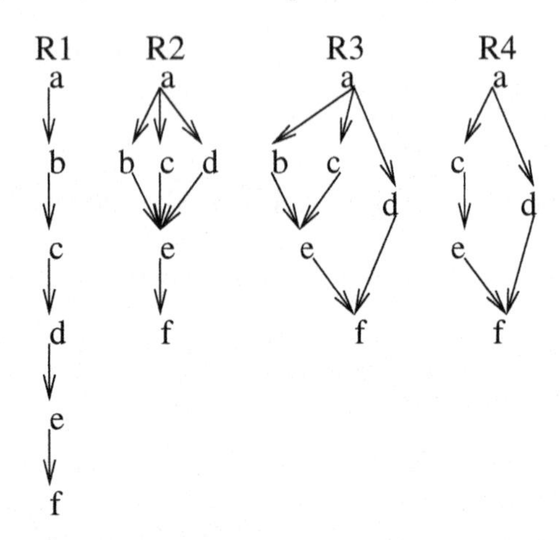

Fig. 5.3. Four orders $R_1 \supset R_2 \supset R_3 \supset R_4$.

instead of general sequences, and assume that each item appears in a string at most once, but not necessarily every item appears in a string.

A string can be written as $s = x_1 \cdots x_l$, where $x_1, \ldots, x_l \in V$. l is called the *length* of string s, i.e., $len(s) = l$. For strings $s = x_1 \cdots x_l$ and $s' = y_1 \cdots y_m$, s is called a *super-string* of s' and s' a *sub-string* of s if (1) $m \leqslant l$ and (2) there exist integers $1 \leqslant i_1 < \cdots < i_m \leqslant l$ such that $x_{i_j} = y_j$ ($1 \leqslant j \leqslant m$). We also say s contains s'. For a string database SDB, the *support* of a string s, denoted by $sup(s)$, is the number of strings in SDB that are super-strings of s.

The total order defined by string s can be written in the *transitive closure* of s, denoted by $\mathcal{C}(s) = \{(x_i, x_j) | 1 \leqslant i < j \leqslant l\}$. Please note that, in the transitive closure, we omit the trivial pairs (x_i, x_i). For example, for string $s = abcd$, $len(s) = 4$. The transitive closure is $\mathcal{C}(s) = \{(a,b), (a,c), (a,d), (b,c), (b,d), (c,d)\}$. Here, we omit the trivial pairs (a,a), (b,b), (c,c), and (d,d).

The *order containment relation* is defined as, for two partial orders R_1 and R_2, if $R_1 \subset R_2$, then R_1 is said to be *weaker* than R_2 and R_2 is *stronger* than R_1. By intuition, a partially ordered set (or poset for short) satisfying R_2 will also satisfy R_1. For example, in Figure 5.3, $R_4 \subset R_3 \subset R_2 \subset R_1$. Please note that R_4 covers fewer items than the other three partial orders. Trivially, we can add the missing items into the DAG as isolated vertices so that every DAG covers the same set of items. To keep the DAG simple and easy to read, we omit such isolated items.

Frequent Closed Partial Orders (FCPO)

A *string database* SDB is a multiset of strings. For a partial order R, a string s is said to *support* R if $R \subseteq \mathcal{C}(s)$. The *support of* R *in* SDB, denoted by $sup(R)$, is the number of strings in SDB that support R. Given a minimum support threshold min_sup, a partial order R is called *frequent* if $sup(R) \geqslant min\_sup$.

Following the related definitions and the order containment relation, we have the following result.

Property 5.2 (Anti-monotonicity of frequent partial orders). For a string database SDB and partial orders R and R' such that $R' \subset R$, it is the case that $sup(R') \geqslant sup(R)$. Therefore, if R is frequent, then R' is also frequent. ∎

To avoid the triviality, instead of reporting all frequent partial orders, we can mine the representative ones only.

Example 5.3 (Frequent closed partial orders). Let us consider string database DB in Table 5.2 again. The four sequential patterns discussed in Example 5.1 can be regarded as frequent partial orders which are supported by strings 1, 2 and 4. As discussed before, given that the partial order R in Figure 5.1 is also supported by strings 1, 2 and 4, the four sequential patterns as frequent partial orders are redundant.

There does not exist another partial order R' such that R' is stronger than R in Figure 5.1 and is also supported by strings 1, 2 and 4. In other words, R is the strongest one among all frequent partial orders supported by strings 1, 2 and 4. Thus, the partial order R is not redundant and can be used as the representative of the frequent partial orders supported by strings 1, 2 and 4. Technically, R is a frequent closed partial order. ∎

A partial order R is *closed* in a string database SDB if there exists no partial order $R' \supset R$ such that $sup(R) = sup(R')$. A partial order R is a *frequent closed partial order* if it is both frequent and closed.

Problem Definition. The problem of *mining frequent closed partial orders from strings* is to find the complete set of frequent closed partial orders in a given string database SDB with respect to a minimum support threshold min_sup. ∎

Various Types of Frequent Patterns from Strings

For a string database SDB and a minimum support threshold min_sup, in addition to frequent closed partial orders, we can mine some other types of frequent patterns as follows.

Frequent itemsets [2] and frequent closed itemsets [84] . If the ordering information in a string is ignored, a string can be treated as a set of items. For a set of items $X \subseteq I$, $sup(X)$ is the number of strings in SDB in which X appears. X is a *frequent itemset* if $sup(X) \geqslant min\_sup$. A frequent itemset X is a *frequent closed itemset* if there exists no $X' \supset X$ such that $sup(X) = sup(X')$.

Sequential patterns [3] and closed sequential patterns [125]. For string s, $sup(s)$ is the number of strings in SDB which contain s as a substring. s is a *sequential pattern* if $sup(s) \geqslant min\_sup$. In other words, a sequential pattern is a frequent total order on a subset of items. A sequential pattern s is a *closed sequential pattern* if there exists no proper super-sequence s' of s such that $sup(s') = sup(s)$.

Graph patterns [53] and closed graph patterns [124]. Since a string defines a total order, its transitive closure can be viewed as a DAG. For a DAG G, $sup(G)$ is the number of graphs in which G is an embedded subgraph. G is a *frequent graph pattern* if $sup(G) \geqslant min\_sup$. A frequent graph pattern G is a *frequent closed graph pattern* if there exists no graph G' such that $sup(G) = sup(G')$ and G is an embedded subgraph of G'.

Then, *what are the relationships among the above types of frequent patterns?* We have the following results based on the related definitions.

Corollary 5.4 (FPO, frequent itemsets and sequential patterns). *The set of items in a frequent partial order is a frequent itemset. Moreover, if R is a frequent partial order, then, every path s in R is a sequential pattern.* ∎

In a directed acyclic graph (DAG), a vertex v is a *sink* if no edge leaves v. A vertex v is a *source* if no edge enters v. The relationships among frequent closed partial orders, closed sequential patterns and transitive closure graph patterns are described in the following theorems. We leave the proof to the interested readers as an exercise.

Theorem 5.5 (FCPO and closed sequential patterns). *Let R be a frequent closed partial order, and $s_1, \ldots, s_k$ be all the paths in R's transitive closure graph such that each path is from a source to a sink. Then, a string s supports R if and only if it simultaneously supports $s_1, \ldots, s_k$.* ∎

Theorem 5.6 (CPO and transitive closure graph patterns). *A partial order R is a frequent (closed) partial order in a string database if and only if it is a frequent (closed) graph pattern in the corresponding database of transitive closures of strings.* ∎

5.1.2 How Is Frequent Closed Partial Order Mining Different from Other Data Mining Tasks?

Sequential Pattern Mining

As sequence data is available in many applications, mining sequence data has been investigated extensively. There are intensive studies on mining sequential patterns, as described in Chapter 2. Several efficient algorithms were proposed, such as GSP [101], PrefixSpan [88], SPADE [132], SPAM [5], and DISC [19]. There can be many sequential patterns. To improve the effectiveness and remove the redundant sequential patterns, closed sequential patterns can be mined [125, 114, 109]. Another approach to improve the effectiveness is to specify constraints. In [34, 89], various constraints on sequential patterns were investigated.

In [76], Mannila et al. considered mining frequent episodes from event sequences (basically, strings). In principle, an episode can be any partial order. However, due to the computational complexity consideration, algorithms on only series and parallel episodes were given. An episode is parallel if the partial order is trivial (i.e., $x \nleqslant y$ for all $x \neq y$). An episode is series if the partial order is a total order (i.e., for any x and y, either $x \leqslant y$ or $y \leqslant x$). It coincides with sequential pattern mining in general.

As illustrated in Example 5.1, mining frequent partial orders is a generalization of mining sequential patterns.

Mining Partial Orders

Recently, two interesting studies investigated the problem of mining a small set of partial orders globally fitting data best [74, 37]. Particularly, [74] addressed sequence data. However, very different from the problem studied here, [74] tried to find one or a (small) set of partial orders that fit the *whole* data set as well as possible, which is an optimization problem. An implicit assumption is that the whole data set somehow follows a global order. We will discuss the global order mining problem in Section 5.2.

Moreover, [1, 112] studied the problem of reconstructing a workflow model from a set of executions of the model, such as records in a log file. In process model mining, it is also assumed that a global workflow template exists and the mining wants to reconstruct the template as much as possible from the executions of the template. The assumption of a global template is feasible and useful in some applications, such as scheduling jobs and students taking courses.

For some other applications, such as the DNA microarray data analysis and network packet routing, there is usually no non-trivial order that can be expected globally. This chapter addresses such situations. That is, we want to find the partial orders that are frequent in a database, but not necessarily dominate the database. Some partial orders found may even conflict with each other.

There is another important difference between the work [74] and the problem discussed in this chapter. In [74], due to the complexity consideration, only series-parallel orders [110] are considered, whose definition is recalled as follows.

The *minimal series parallel* (MSP) DAG is defined as follows.

1. The DAG having a single vertex and no edges is MSP;
2. If $G_1 = (V_1, E_1)$ and $G_2 = (V_2, E_2)$ are two MSP DAGs, so is either of the DAGs constructed by the following operations.
   a) *Parallel composition*: $G_p = (V_1 \cup V_2, E_1 \cup E_2)$;

b) *Series composition:* $G_s = (V_1 \cup V_2, E_1 \cup E_2 \cup (N_1 \times R_2))$, where N_1 is the set of sinks of G_1 and R_2 is the set of sources of G_2.

A partial order can be represented as a DAG. A partial order is a *series parallel order* if the transitive reduction of its DAG is an MSP DAG.

Intuitively, a series parallel order is formed by assembling objects using parallelism and serialism. An important property is that a series parallel order can be represented in a *binary decomposition tree* [110]. Then, many search problems can be solved efficiently by dynamic programming.

In [74], Mannila and Meek tried to find series parallel orders that globally fit a data set as well as possible. However, for mining frequent partial orders, series parallel orders may not be sufficient, since they cannot always capture all the partial orders shared by sequences.

Example 5.7 (Dimension 2 non-series parallel order). Consider two strings *abdc* and *dacb*. A partial order R shared by them is shown in Figure 5.4. Since R is exactly the forbidden subgraph of MSP DAG [110], R is not series parallel. In other words, only using the series parallel orders cannot cover all frequent partial orders. ■

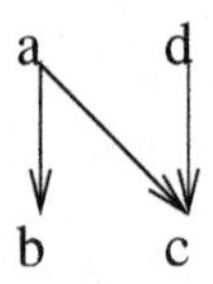

Fig. 5.4. The frequent partial order R shared by *abdc* and *dacb*.

For an order R, the *dimensionality* is defined as the minimum number of total orders whose intersection is R. It is shown [110] that any series parallel order is the intersection of two total orders, that is, the dimensionality of any series parallel order is 2. For partial orders that are frequent in multiple sequences (i.e., they are the intersection of the corresponding total or partial orders), the dimensionality is likely more than two. In such cases, the frequent partial orders may not be series parallel.

We can find non-series parallel orders that are frequent in real data sets. Figure 5.5 presents some examples (except for Figure 5.5(b)). Such non-series parallel orders cannot be identified by any previous methods.

There are several interesting studies on applications of ordering information. For example, to discover local structures in gene expression data, Ben-Dor et al. [8] looked for local patterns that manifest themselves simultaneously on a subset G of genes and a subset T of experiments. Specifically, they searched for order-preserving submatrices (OPSMs), in which the expression levels of all genes induce the same linear ordering (i.e., total order on a subset) of the experiments. They showed that the OPSM search problem is NP-hard.

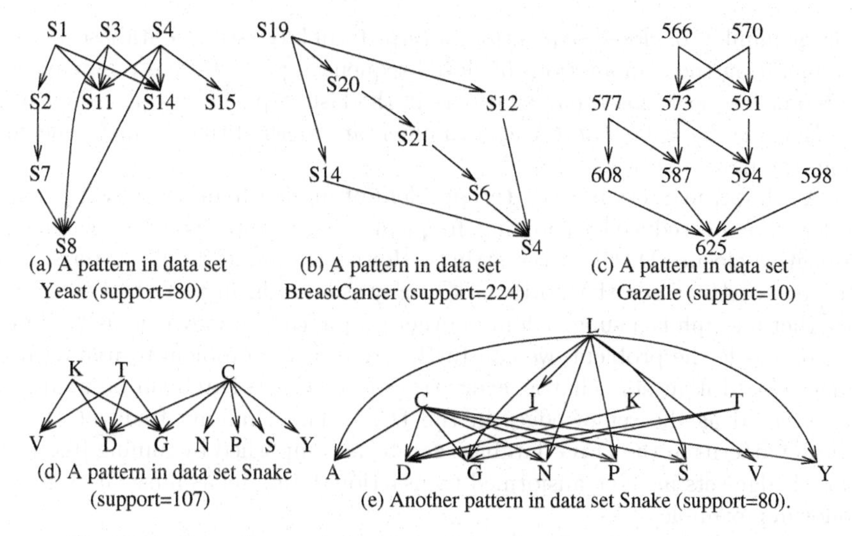

(a) A pattern in data set Yeast (support=80)
(b) A pattern in data set BreastCancer (support=224)
(c) A pattern in data set Gazelle (support=10)
(d) A pattern in data set Snake (support=107)
(e) Another pattern in data set Snake (support=80).

Fig. 5.5. Some frequent closed partial orders found in real data sets.

They defined a probabilistic model in which an OPSM is hidden within an otherwise random matrix. Guided by this model, they developed an efficient algorithm to find hidden OPSMs in a random matrix. Please note that their method cannot find the *complete set* of linear orders. Instead, our methods here find the complete set. In [68], Liu and Wang proposed a sequential pattern mining method to find the complete set of linear orders, i.e., substrings. However, their approach is not concerned with partial orders in general.

5.1.3 TranClose: A Rudimentary Method

The relationship between frequent closed partial orders and other types of frequent patterns from string databases (Section 5.1.1) suggests some rudimentary methods to mine frequent closed partial orders by reducing the problem to mining other types of frequent patterns. For example, a naïve 3-step method is as follows.

First, we mine the complete set of closed sequential patterns. Then, we enumerate the combinations of closed sequential patterns. For a set of closed sequential patterns $s_1, \ldots, s_n$, let R be a partial order generated as follows. $(x, y) \in R$ if and only if (1) $x \neq y$; (2) there exists a sequential pattern s_i $(1 \leqslant i \leqslant n)$ such that x appears before y in s_i; and (3) y never appears before x in any sequential pattern s_j $(1 \leqslant j \leqslant n)$. If $R \neq \emptyset$ then a closed partial order is identified. Last, we check the support of the closed partial orders identified in the previous step, and remove all redundant frequent closed partial orders. That is, a frequent closed partial order should be output only once.

However, such a naïve algorithm is far from efficient. First, mining the complete set of closed sequential patterns is non-trivial. Second, there often exist a large number of closed sequential patterns from large string databases. Last, enumerating the combinations of closed sequential patterns and removing redundant frequent closed partial orders in the last step can be very expensive.

Here, we describe *TranClose*, a rudimentary method that is more efficient than the naïve method.

As shown in Section 5.1.1, the problem of mining frequent closed partial orders can be reduced to mining frequent closed graph patterns from the transitive closure DAGs of the strings. However, mining graph patterns can be very costly, since the bottleneck, many isomorphism tests to determine whether a graph is a subgraph in another graph, can be very expensive [124].

To tackle the problem, we can further reduce the problem to mining frequent closed itemsets. That is, every transitive closure DAG can be uniquely represented as the set of edges in the DAG. Then, mining frequent closed graph patterns in the DAG database can be accomplished by mining frequent closed edge-sets in the transformed transaction database, as illustrated in the following example.

Example 5.8 (TranClose). Consider a string database SDB as shown in the first two columns of Table 5.3. Suppose the minimum support threshold is 2.

Sid	String	Transitive closure $\mathcal{C}(s)$
1	$abcdef$	$ab, ac, ad, ae, af, bc, bd, be, bf, cd, ce, cf, de, df, ef$
2	$acbde$	$ac, ab, ad, ae, cb, cd, ce, bd, be, de$
3	$dabce$	$da, db, dc, de, ab, ac, ae, bc, be, ce$
4	$dcabe$	$dc, da, db, de, ca, cb, ce, ab, ae, be$

Table 5.3. String database SDB as the running example.

TranClose mines the complete set of frequent closed partial orders in three steps.

In the first step, we expand the strings to their transitive closures. A transitive closure is denoted by the set of edges. The third column of Table 5.3 shows the transformation. The set of edges in the transitive closure of each string becomes a transaction so a transaction database TDB is created. Please note that the edges are directed. For example, edges bc and cb are different.

In the second step, we mine frequent closed edge-sets from the transformed transaction database TDB (i.e., the third column in Table 5.3) with support threshold min_sup = 2. Each frequent closed edge-set corresponds to the transitive closure of a frequent closed partial order.

In the last step, for each frequent closed edge-set, we compute its transitive reduction. In this example, there are six patterns found. They are shown in Figure 5.6 together with their transitive reduction DAGs for examination. ■

Algorithm *TranClose* is summarized in Figure 5.7. We can use any frequent closed itemset mining algorithm such as CHARM [133] and CLOSET+ [115]

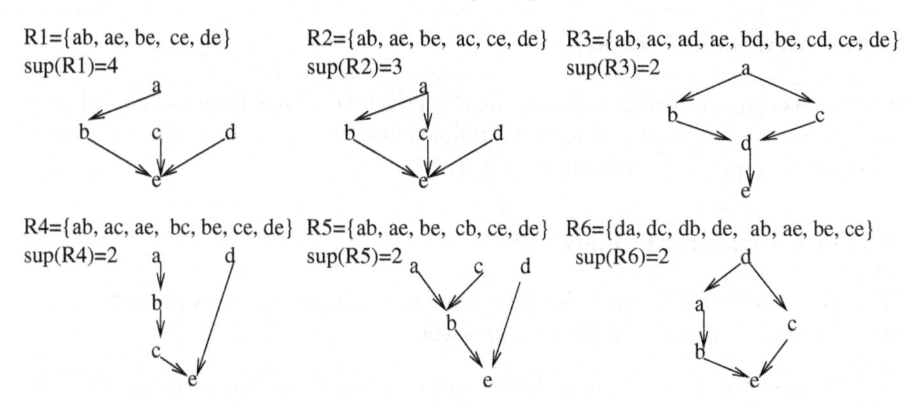

Fig. 5.6. The frequent closed itemsets and the transitive reductions of the corresponding frequent CPOs.

to mine frequent closed edge-sets from the transformed transaction database, and derive the frequent closed partial orders from the frequent closed edge-sets.

Input: a string database SDB and a minimum support threshold min_sup;
Output: the complete set of frequent closed partial orders;
Method:
1: create a transaction database TDB by transforming each string in SDB into its transitive closure and make the set of edges in the transitive closure a transaction;
2: mine frequent closed edge-sets from TDB with support threshold min_sup;
3: for each frequent closed edge-set, compute its transitive reduction as a frequent closed partial order;

Fig. 5.7. The *TranClose* algorithm.

The bottleneck of *TranClose* is that it has to handle a very much enlarged transitive closure database: for a string of length l, its transitive closure has $\frac{l(l-1)}{2}$ edges.

In our small running example, there are totally 41 edges in all the frequent closed edge-sets, while only 27 edges in their transitive reduction DAGs. In other words, more than one-third of the edges in the transitive closures are redundant and will be removed in the transitive reductions. For large string databases where there are long strings, the redundancy may be even bigger. As a well accepted fact, mining long patterns is often very costly. *To improve the efficiency, we have to avoid computing transitive closure in mining frequent closed partial orders.*

5.1.4 Algorithm Frecpo

In this section, we present algorithm *Frecpo* which mines frequent closed partial orders in the form of transitive reductions directly from string databases and avoids computing transitive closures.

General Idea and Framework

In order to efficiently mine the complete set of frequent closed partial orders, we have to address the following two issues.

- *The correctness and completeness issue.* We have to find a systematic way to enumerate all the frequent closed partial orders without duplicate. This will guarantee that the mining result is correct and complete.
- *The efficiency and scalability issue.* We must have an efficient method to extract frequent closed partial orders and prune futile search branches.

To address the correctness and completeness issue, *Frecpo* searches a set enumeration tree of transitive reductions of partial orders in a depth-first manner.

In principle, a partial order can be uniquely represented as the set of edges in its transitive reduction. Moreover, all edges in a set can be sorted in the dictionary order[1] and thus it can be written as a list. Therefore, we can enumerate all partial orders in the dictionary order. A set enumeration tree of partial orders can be formed: for orders R_1 and R_2, R_1 is an ancestor of R_2 and R_2 is a descendant of R_1 in the tree if and only if the list of edges in R_1 is a prefix of the list of edges in R_2.

For example, consider a set of items $\{a, b, c\}$. The transitive reductions of all possible partial orders on the three items can be enumerated in a set enumeration tree shown in Figure 5.8.

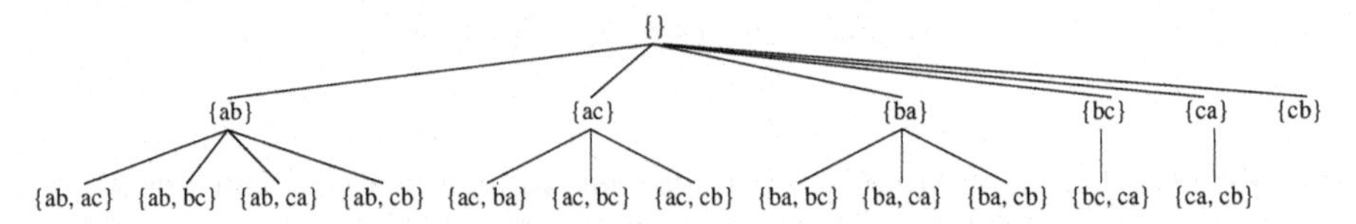

Fig. 5.8. The set enumeration tree of the transitive reductions of all possible partial orders on items a, b and c.

By a depth-first search of the set enumeration tree of transitive reductions of partial orders, *Frecpo* will not miss any frequent partial order. *Frecpo* employs depth-first search instead of breadth-first search because there are many

[1] In fact, any global order on the edges works. For the sake of convenience, we choose dictionary order as an example here.

previous studies (e.g., [19, 68, 88, 115, 124, 125, 133, 132]) strongly suggesting that a depth-first search with appropriate pseudo-projection techniques often achieves a better performance than a breadth-first search when mining large databases.

To address the efficiency and scalability issue, *Frecpo* prunes the futile branches and narrows the search space as much as possible. Basically, three types of techniques are used.

- *Pruning infrequent items, edges and partial orders.* According to Property 5.2, if a partial order R in the set enumeration tree is infrequent, then the partial orders in the subtree rooted at R, which are stronger than R, cannot be frequent. The subtree can be pruned. Hence, *Frecpo* often does not have to search the complete set enumeration tree. Instead, only the upper part of the tree which contains all the frequent partial orders is searched. Moreover, only frequent closed partial orders will be output.
- *Pruning forbidden edges.* Not every edge can appear in the transitive reduction of a partial order. For example, if every string containing ac also contains ab and bc, then edge ac should not appear in the transitive reduction of any frequent closed partial order. Edge ac is called a forbidden edge. Removing the forbidden edges can also reduce the search space.
- *Extracting transitive reductions of frequent partial orders directly.* In *Frecpo*, we develop an efficient method to identify frequent closed partial orders and also extract their transitive reductions from various subsets of strings. Thus, *Frecpo* does not need to compute the transitive reductions.

Input: a string database SDB and a minimum support threshold *min_sup*;
Output: the complete set of frequent CPOs;
Method:

1: scan database once, find frequent items; // Lemma 5.9
2: scan database again, find global feasible edges; // Lemmas 5.9 and 5.10
   // if the number of items in SDB is not large, the first two scans can
   // be combined.
3: let R be the set of global feasible edges with support $|SDB|$;
4: if $R \neq \emptyset$ then output R as a frequent CPO; // Lemma 5.13
5: let $L = e_1, \ldots, e_n$ be the list of global feasible edges with support less than $|SDB|$;
6: for each edge e_i in L do
7:    if $R \cup \{e_i\}$ does not contain any redundant edge and there exists no FCPO R' found before such that $R' \supset (R \cup \{e_i\})$ and $sup(R') = sup(e_i)$ then
8:       form $R \cup \{e_i\}$-projected database $SDB|_{R\cup\{e_i\}}$;
9:       recursively mine $SDB|_{R\cup\{e_i\}}$;

Fig. 5.9. The *Frecpo* algorithm.

Algorithm *Frecpo* is shown in Figure 5.9. In the following subsections, we will explain the technical details.

Pruning in *Frecpo*

The first rule of pruning is a corollary of Property 5.2.

Lemma 5.9 (Pruning by support). *An infrequent item or an infrequent edge cannot appear in any frequent partial order.* ∎

The lemma is used in *Frecpo* in two ways. First, at the beginning of the algorithm, the database is scanned so that frequent items and frequent edges are identified. Infrequent items and infrequent edges are pruned. Second, in the recursive depth-first search, for any frequent closed partial order R, only the edges that frequently appear together with R in the string database should be used to expand R to form R's children. Technically, all the strings supporting R form the R-projected database $SDB|_R = \{s \in SDB | R \subseteq \mathcal{C}(s)\}$. Only the frequent edges in the R-projected database and satisfying the requirement of enumeration tree should be used to expand R to R's children in the set enumeration tree. The items and edges infrequent in the projected database will be removed.

The second rule of pruning is based on the observation that not every frequent edge can appear in the transitive reduction of a frequent closed partial order. An edge xy is called a *forbidden edge* in a string database SDB if there exists an item z such that for every string s in SDB which contains xy, s also contains xzy. In such a case, for any frequent closed partial order R which contains (x, y), R also contains (x, z) and (z, y), which disqualify (x, y) in R's transitive reduction.

Lemma 5.10 (Pruning forbidden edges). *A forbidden edge cannot appear in the transitive reduction of any frequent closed partial order.* ∎

Frecpo uses a *detection matrix* to identify both frequent edges and forbidden edges, as illustrated in the following example.

*Example 5.11 (*Frecpo *– Part 1).* Let us consider again mining frequent closed partial orders from the string database SDB in Table 5.3 with respect to minimum support threshold $min\_sup = 2$.

By scanning the database only once, *Frecpo* computes the supports of the items. Following Lemma 5.9, infrequent items are pruned, such as f in our running example ($sup(f) = 1$).

To prune the infrequent edges and the forbidden edges, *Frecpo* scans the database again and fills in a matrix $\{cnt[x, y]\}$, where x and y are both frequent items, and $cnt[x, y]$ registers both $sup(xy)$ and the list of items that appear between x and y in all strings having been scanned so far that contain xy. The list is called the anchor list. The matrix is called the *detection matrix* and is shown in Figure 5.10.

	a	b	c	d	e
a		$4, \emptyset$	$3, \emptyset$	$2, \{b, c\}$	$4, \{b\}$
b	$0, \emptyset$		$2, \emptyset$	$2, \emptyset$	$4, \emptyset$
c	$1, \emptyset$	$2, \emptyset$		$2, \emptyset$	$4, \emptyset$
d	$2, \emptyset$	$2, \{a\}$	$2, \emptyset$		$4, \emptyset$
e	$0, \emptyset$	$0, \emptyset$	$0, \emptyset$	$0, \emptyset$	

Fig. 5.10. The matrix detecting infrequent edges and edges not in transitive reduction.

From the detection matrix, we can immediately prune the infrequent edges (those with support less than 2, such as ca). An edge is a forbidden edge if its anchor list is not empty. Following Lemma 5.10, forbidden edges ad, ae and db can be pruned as well.

In this example, SDB contains 6 different items. There are $6 \times 5 = 30$ possible different edges. 20 different edges appear in the database. Only 11 edges survive from the pruning. ∎

Clearly, the length of an anchor list monotonically decreases as the scan goes on. If the strings are scanned in an arbitrary order, the initial length of any anchor list for any edge is bounded by the maximum number of frequent items in any string. More often than not, the length of a string is much shorter than the total number of items in the whole database. Moreover, as a heuristic, we can scan the short strings before the long ones. Then, the initial length of any anchor list for any edge xy is bounded by the number of frequent items in the shortest string which contains xy.

If the number of items is not very large and a detection matrix for all items can be held in main memory, *Frecpo* can scan the database only once to prune infrequent items, infrequent edges and forbidden edges by using a detection matrix holding all items instead of only the frequent ones.

Extracting Frequent Closed Partial Orders

*Example 5.12 (*Frecpo *– Part 2).* As shown in Example 5.11, only the edges ab, ac, bc, bd, be, cb, cd, ce, da, dc and de can be used to construct the transitive reduction of a frequent closed partial order. They are called the *global feasible edges*. Among them, ab, be, ce and de have support 4, i.e., they appear in every string in SDB. The four edges form a frequent closed partial order, i.e., order R_1 in Figure 5.6. In other words, the set of global feasible edges that appear in every string forms a frequent CPO. Interestingly, the set is in fact the transitive reduction, since any redundant edge in the set is identified as a forbidden edge by the detection matrix.

The observation in Example 5.12 leads to the following.

Lemma 5.13. (EXTRACTING TRANSITIVE REDUCTION OF FCPO). *In a string database SDB, the set of global feasible edges that have support $|SDB|$ is the transitive reduction of the frequent closed partial order R of support $|SDB|$.*

Proof. There exists only one FCPO whose support is $|SDB|$. Otherwise, if there are two FCPOs R_1 and R_2 whose support are $|SDB|$, both R_1 and R_2 ($R_1 \neq R_2$) are supported by every string in the database. That means $R_1 \cup R_2$ is supported by every string in the database and thus is also a frequent partial order with support $|SDB|$. That leads to a contradiction to the assumption that R_1 and R_2 are closed.

We denote the FCPO with support $|SDB|$ by R. The set specified in the lemma is a superset of the transitive reduction of R. On the other hand, if there exists a redundant edge, the edge will be identified by the detection matrix. Hence, the set is exactly the transitive reduction of R. ■

Lemma 5.13 enables *Frecpo* to identify the transitive reduction of frequent partial orders directly. In other words, *Frecpo* prunes redundant edges using detection matrix. It never has to explicitly compute transitive reduction for any frequent closed partial order.

Recursively Depth-first Searching

Once a frequent closed partial order R is found, *Frecpo* expands R to its children. Following the similar reasoning in Lemmas 5.9, 5.10 and 5.13, only frequent, non-forbidden edges in the R-projected database should be used to expand R to its children in the enumeration tree.

*Example 5.14 (*Frecpo *– Part 3).* Let us continue the mining process in Example 5.12. Frequent closed partial order R_1 is the order shared by all strings. Thus, any other frequent closed partial order will be stronger than R_1.

The other frequent closed partial orders in transitive reduction can be partitioned into the following subsets according to the dictionary order of the remaining global feasible edges (i.e., *ac*, *bc*, *bd*, *cb*, *cd*, *da*, and *dc*): (1) the ones having edge *ac* in their transitive reduction; (2) the ones having edge *bc* but no *ac* in their transitive reduction; ...; and (7) the one having *dc* but no other edges in its transitive reduction (if it is a frequent closed partial order). These subsets can be mined one by one in a depth-first search manner.

We first consider the subset of frequent closed partial orders having edge *ac* in their transitive reductions. They also contain R_1. The strings in SDB that are super-strings of *ac*, namely strings 1, 2 and 3, are collected as the $(R_1 \cup \{ac\})$-projected database.

We prune the local infrequent items, infrequent edges and forbidden edges by scanning the $(R_1 \cup \{ac\})$-projected database once and filling in the local detection matrix. The feasible edges in this projected database are *bc*, *bd* and *cd*. Since each feasible edge has support 2, which is less than the number of strings in the projected database, we extract $R_2 = R_1 \cup \{ac\}$ as the transitive reduction of a frequent closed partial order as shown in Figure 5.6. Any frequent partial order having ac must be stronger than R_2.

Since we have three local feasible edges in the $(R_1 \cup \{ac\})$-projected database, the remaining frequent closed partial orders having ac in their transitive reduction can be further partitioned into three sub-subsets: the ones having ac and bc, the ones having ac and bd but no bc, and the ones having ac and cd but no bc nor bd.

R_2 has an edge ab, and any frequent partial order having ac is a superset of R_2. Clearly, edges ab, ac and bc cannot stay together in a transitive reduction, since ac is redundant in such a case. Thus, we immediately determine that the first sub-subset is empty without checking the database at all.

The remaining frequent closed partial orders can be found recursively. ∎

Summary

In implementation, the pseudo-projection technique can be used, which was firstly proposed in [88] and later has become popular in depth-first search frequent pattern mining. That is, if a database or a projected database can fit into main memory, instead of deriving a copy of strings for every projected database, we use hyperlinks (implemented as pointers) to link the strings in the projected database together. The recursively projected databases can share the same physical database storage. Scanning and deriving projected databases are efficient with the help of hyperlinks. As discussed in [88, 114], if a database is large and cannot fit into main memory, the physical projections should be generated. Once a projected database can be held into main memory, the recursion is switched to pseudo-projection.

The correctness of algorithm *Frecpo* can be justified based on our previous discussion. Comparing to algorithm *TranClose* and other rudimentary methods, *Frecpo* has three distinct advantages.

Advantage 1: Mining in transitive reduction to avoid substantial space and I/O overhead. *Frecpo* never explicitly unfolds strings into transitive closures. As discussed before, the transitive closures of strings can be much larger than the strings themselves. Thus, mining the strings directly avoids the substantial space overhead and also the I/O cost. *Frecpo* does examine combinations of items for each string. However, such tests are conducted on the fly in main memory. It does not involve any space or I/O overhead, which is the bottleneck of mining large databases.

Advantage 2: Directly extracting frequent closed partial orders in transitive reduction. *Frecpo* computes detection matrices and extracts frequent closed partial orders in transitive reduction directly (Lemma 5.13). It avoids the post-processing of computing transitive reductions. Transitive reduction is the minimum representation of a partial order. Using this minimum representation makes the mining more effective and efficient.

Advantage 3: Aggressively and progressively pruning futile branches in recursive depth-first search. *Frecpo* aggressively prunes infrequent items and edges and forbidden edges that are impossible to appear in transitive reductions of frequent closed partial orders. Thus, the search space shrinks dramatically in the recursive depth-first search. Moreover, only frequent items and local feasible edges in the current projected database will be used to expand the current frequent closed partial order into stronger ones. This pattern-growth approach makes the search more focused.

5.1.5 Applications

The knowledge about ordering, especially the frequent partial orders in string databases, has many applications. Here, we list four of them.

Application 1: Bioinformatics. Ordering information is often important in analysis of biological experiment data. For example, to discover patterns in gene expression matrices, one promising approach [8] is to look for order-preserving submatrices (OPSMs). That is, in an n by m gene expression matrix for n genes and m experiments, each element $v_{i,j}$ gives the expression level of a gene g_i in an experiment e_j. A submatrix is order-preserving if the expression levels of all genes in the submatrix induce the same (linear or partial) ordering of the experiments. As indicated in [8], such a pattern may arise if the experiments in the order-preserving submatrix represent distinct stages in the progress of a disease or in a cellular process, and the expression levels of all genes in the submatrix vary across the stages in the same way. Moreover, [68] also shows that a partial order of conditions shared by a group of genes may indicate that the genes form a co-expressed group and they respond to a sequence of environment stimuli.

Application 2: Process model mining, web mining and market basket analysis. The workflow paradigm has been extensively used to specify how business processes should be conducted. It is often desirable to construct process models from logs of past, unstructured executions of a given process [1, 112].

In web mining and market basket analysis, a critical task is to identify groups of customers in which all customers' sequences of purchases induce the same ordering of a series of products. In previous studies, sequential patterns are often used for this purpose. However, as shown in Example 5.1, sequential patterns may not be able to concisely capture the general ordering information. Instead, a partial order can model the customers' purchase behavior better. Thus, it can be more informative and more effective to use frequent partial orders in place of sequential patterns in many cases. Moreover, selected frequent partial orders can be used as signatures of customer behavior in classification and clustering analysis.

Application 3: Network management and intrusion detection. In network management, it is important to characterize network traffic. Frequent partial orders obtained from network packet scheduling data may disclose frequent routing paths and identify possible bottlenecks of networks.

Moreover, it is important to discover the signatures (i.e., distinct features) of normal network access and intrusions. Consider misuse detection, where a training data set containing both labeled normal activities and intrusions is available. We can mine from the training data set partial orders which are frequent in the subset of intrusions and are rare in the subset of normal activities. Such frequent partial orders can be used to identify malicious activities in the future. On the other hand, in anomaly detection, frequent partial orders can be used to characterize the major patterns of network accesses. If an activity does not follow any frequent partial orders observed so far, then it can be a candidate of anomaly.

Application 4: Preference-based service. Preferences can be modeled as partial orders. It is interesting to study common preferences from a large collection of data, such as marketing survey and product evaluation. For example, a customer may be asked to rank a set of products in a marketing survey. The preference of a customer can be derived from her/his ranking. Then, it is interesting to mine the common preferences as frequent partial orders from the ranking data. Moreover, customer segmentation and marketing campaigns may be developed based on such ordering information.

5.2 Mining Global Partial Orders

In Section 5.1, we discussed how to find partial orders that are frequent in a sequence database. Generally, we may find many frequent (closed) partial orders in a sequence database. In some applications, we may want to find one partial order that fits all the sequences as well as possible. This problem is called the global partial order mining problem, and was first systematically treated in [74].

5.2.1 Motivation and Preliminaries

Let us consider the sequences in Table 5.2 again. If we want to find a partial order to summarize all sequences in the database, which order should be returned?

As discussed in Example 5.1, the partial order in Figure 5.1 fits sequences 1, 2, and 4 nicely. The order does not fit some segments of sequence 3. That is, Data Structure is taken before Programming, and Software Engineering is taken after Data Mining and Information Retrieval in sequence 3.

Overall, the order in Figure 5.1 is the best to summarize the whole set of 4 sequences.

How can we determine whether a partial order fits a sequence?

Let V be the alphabet set of our sequence database. For a partial order R and a sequence s, s is compatible with R if for any events $u, v \in V$, $(u, v) \in R$

then $uv \in s$. s is an extension of R if s is compatible with R and s is a total order—that is, s is a string. Moreover, s is a complete extension of R if s is an extension of R and s contains all events of R.

Given a partial order R, how many complete extensions are there for R? If R is a total order, then R has only one complete extension. However, if R is a trivial order, that is, for any $u \neq v$, neither (u, v) nor (v, u) are in R, then R has $|V|!$ complete extensions.

When we try to find a partial order R to fit a set of sequences, we want, on the one hand, as many sequences as possible are compatible with R; on the other hand, R is as strong as possible (that is, more ordering information is retained). For example, in the sequence database in Table 5.2, a trivial order fits every sequence, but does not contain any ordering information. It is too general. On the other hand, any total order may fit at most one sequence completely, and thus cannot represent other sequences well. That is, a total order is too specific. The order in Figure 5.1 is a nice tradeoff between the generality and the specificity.

Mathematically, let $\alpha(R)$ be the number of complete extensions of R. For a sequence s that is compatible with R, the probability of s given R is $P(s|R) = \frac{1}{\alpha(R)}$. If s is not compatible with R then $P(s|R) = 0$.

Let us assume that the sequences in a sequence database S are independently and identically distributed. Then, the probability of the set of sequences S given the partial order R is

$$P(S|R) = \prod_{s \in S} P(s|R)$$

Problem definition. Given a sequence database S, the problem of mining a global partial order is to find partial order R_g such that R_g maximizes $P(S|R)$. That is, $R_g = \arg\max_{\text{partial orders } R}\{P(S|R)\}$.

5.2.2 Mining Algorithms

A critical issue in mining a global partial order is to compute $\alpha(s|R)$ for a sequence s and a partial order R. Unfortunately, this problem has been shown NP-hard [11][2]. In [74], an efficient method was developed to mine global partial orders in the form or series-parallel partial orders (definition in Section 5.1.2).

[2] In fact, it is #P-complete. #P-Complete problems [111, 96] are the enumeration problems which might be intractable even if P=NP. A typical example of #P-Complete problem is counting the number of Hamiltonian circuits in a graph. Even if P=NP (i.e., we could tell in polynomial time whether an arbitrary graph contains a Hamiltonian circuit), it is unclear that this would enable us to count the number of Hamiltonian circuits in an arbitrary graph in polynomial time. However, not every, enumeration problem is #P-Complete. For example, there are polynomial time algorithms for the problems of counting the number of Eulerian paths and counting the number of distinct spanning trees in an arbitrary graph.

Using Construction Trees of Series-Parallel Partial Orders

A series-parallel partial order R can be represented as a binary construction tree T_R. Each leaf node in the tree is an event in R, and each internal node in the tree is either a series combination (an S-node) or a parallel construction (a P-node). For an S-node, let u be a leaf node in the left subtree, and v be a leaf node in the right subtree. $(u, v) \in R$. For a P-node, let u be a leaf node in the left subtree and v be a leaf node in the right subtree. Neither $(u, v) \in R$ nor $(v, u) \in R$ hold.

Example 5.15 (Construction tree). Consider the partial order R in Figure 5.11. The construction tree is also shown in the figure.

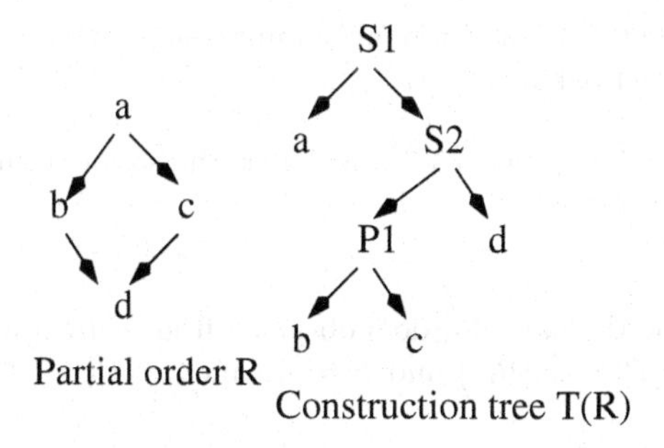

Fig. 5.11. A partial order R and its construction tree T_R.

Each subtree represents a partial order on an exclusive set of events. The P-node $P1$ indicates that neither (b, c) nor (c, b) are in the order. The S-node $S2$ indicates that both (b, d) and (c, d) hold in R. Similarly, we can explain the meaning of S-node $S1$ in the construction tree. ■

For a partial order R, let $n(R)$ be the number of events in R. It is easy to see that if R is a series-parallel partial order, then the construction tree of R can be constructed by scanning the transitive reduction of R once. Moreover, given a sequence s and a series-parallel partial order R, we can test whether s is compatible with R by scanning s and T_R (or the transitive reduction of R) once in a synchronized way. We leave the details of the two algorithms as an exercise for the interested readers.

Using the construction tree of a series-parallel partial order R, we can calculate the number of complete extensions of R as follows.

$$\begin{aligned} \alpha(u) &= 1; \\ \alpha(S(R_1, R_2)) &= \alpha(R_1) \cdot \alpha(R_2); \\ \alpha(P(R_1, R_2)) &= \alpha(R_1) \cdot \alpha(R_2) \cdot \tfrac{n(R_1)+n(R_2))!}{n(R_1)!n(R_2)!}; \end{aligned}$$

where $S(R_1, R_2)$ is an S-node in the tree with R_1 and R_2 as the left and the right subtrees, respectively, $P(R_1, R_2)$ is a P-node in the tree with R_1 and R_2

as the left and the right subtrees, respectively. The computation cost is linear in $n(R)$.

A Greedy Search Method

We can find a global partial order by greedily searching the series-parallel partial orders on the set of possible events. We start with a trivial partial order. Then, we maintain the best order we have got so far, and try to modify the best order to obtain further improvement. The greedy search terminates when the gain in probability is smaller than a threshold in the last iteration. The algorithm is presented in Figure 5.12.

Input: a string database SDB and a quality improvement threshold ϵ;
Output: a series-parallel partial order;
Method:
1: $R =$ the trivial partial order on V, where V is the set of events in R;
2: $p = P(SDB|R)$;
3: `DO`
4: $p_0 = p$, $R_0 = R$;
5: let $R_1, \ldots, R_l$ be the partial orders obtained from R by a modification;
6: let $p = \max_{i=1}^{l}\{P(SDB|R_i)\}$ and $R = \arg\max_{i=1}^{l}\{P(SDB|R_i)\}$;
7: `UNTIL` $(p - p_0) < \epsilon$;
8: `IF` $p > p_0$ `THEN` $R_0 = R$;
9: `RETURN` R_0

Fig. 5.12. The greedy search algorithm for mining the global partial order.

Now, the only problem left is how we can modify the currently best order to generate better orders. This can be done by modifying the binary construction tree of the currently best partial order.

Apparently, in the binary construction tree of a trivial partial order R, all internal nodes are P-nodes. This is the starting point of the greedy search in the algorithm.

Let T_R be the binary construction tree of the currently best partial order R. We generate all possible modifications of R in two steps. In the first step, a leaf node and its parent in T_R are extracted as a small tree. In the second step, the parent-leaf pair is inserted into T_R to generate new modifications by attaching at a new insertion point, choosing a possibly new side (left or right) for the leaf relative to the parent, or changing the type (S-node or P-node) for the parent.

Example 5.16 (Generating modifications). Suppose our greedy search starts with the trivial partial order R as shown in Figure 5.13. Let us consider how to re-arrange c to generate a new order.

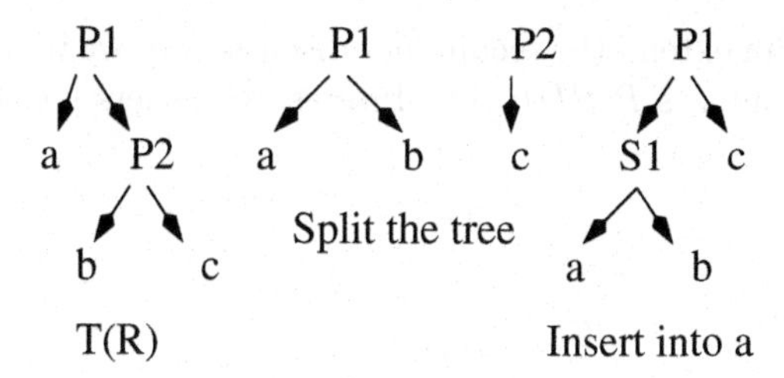

Fig. 5.13. Modifying the currently best partial order.

First, we extract leaf node c and its parent. The tree is split as shown in the figure. In the second step, if the new insertion point is a, we can insert the orphaned tree into a. In the figure, we show one order obtained by inserting c into a as a right subtree of the parent and the parent node type is P-node. By applying the insertion into every possible internal node, and by trying all possible internal node type and both the possible side (left and right), we can generate all possible series-parallel partial orders that can be modified from R.

Suppose R has n events. There are $O(n)$ internal nodes in the construction tree. There are at most $O(n^2)$ possible series-parallel partial orders that can be generated by the modifications. Two modifications may lead to the same series-parallel partial order. Thus, those duplicate orders should be removed.

5.2.3 Mixture Models

We discussed how to mine a series-parallel partial order to globally summarize a set of sequences. However, when there are many conflicting sequences, it is impossible to use one non-trivial series-parallel partial order to fit all sequences. For example, if a database contains three sequences *abcd*, *acbd* and *dbca*, then the only partial order that could summarize the data set is the trivial partial order.

To tackle the cases of conflicting sequences, we can mine a small number of partial orders as the summarization. A mixture model of partial orders consists of k partial orders $R_1, \ldots, R_k$, where k is a parameter. Each order R_i $(1 \leqslant i \leqslant k)$ carries a weight w_i. The probability of a sequence s given a mixture model $\mathcal{M}$ is

$$P(s|\mathcal{M}) = \sum_{R_i \text{ s.t. } s \text{ is compatible with } R_i} \frac{w_i}{\alpha(R_i)}$$

Moreover, the probability of a set of sequences SDB given a mixture model $\mathcal{M}$ is

$$P(SDB|\mathcal{M}) = \prod_{s \in SDB} P(s|\mathcal{M})$$

Generally, we can extend the techniques discussed above to mine a mixture model $\mathcal{M}$ that maximizes $P(SDB|\mathcal{M})$. Interested readers please refer to [74] for details.

5.3 Summary

A major type of useful information that can be discovered from sequence data comes from finding the ordering information hidden in the data. In this chapter, we discussed the techniques to mine ordering information from sequence data. In the first scenario, we want to find frequent closed partial orders in sequences. In the second scenario, we want to find one or a small number of partial orders that fit all the sequences globally.

In practice, the requirements of mining ordering information may vary from one application to another. The techniques described in this chapter provide some typical examples and can be adapted to many situations.

6

Distinguishing Sequence Patterns

A distinguishing sequence pattern is a sequence pattern that (i) characterizes a family of sequences and distinguishes the family from other sequences, (ii) characterizes a special site of sequences and distinguishes the site from other parts of sequences, or (iii) signals something unusual about certain sequences. This chapter first discusses four types of distinguishing sequence patterns, and then gives some methods/algorithms for the mining of two of those types. (The other two types were discussed in Chapter 4.) Distinguishing sequence patterns are also useful as candidates for sequence features.

Three of the types of distinguishing sequence patterns are similar to classification models in the sense that they all discriminate one family of things/sites against some others. There are also differences, since distinguishing sequence patterns are frequently local models (which can match and distinguish some sequences of a class/site) whereas classification models are often total models (which can match and distinguish most sequences of a class/site).

6.1 Categories of Distinguishing Sequence Patterns

There are four types of *distinguishing sequence patterns*, which are summarized in Table 6.1. They are defined based on two factors: whether a special site is considered, and whether one or two datasets are considered. (Recall that a site is a short window of special interest, in the sequences, e.g. transcription binding site, gene start site, or site where certain special events occur. A dataset is usually a family of sequences with certain properties.) When two datasets are considered, we will say one of them is the target dataset and the other is the opposing dataset.

We now describe each of the four types.

- The first type is the *site-characteristic distinguishing sequence patterns*, defined as sequence patterns which are relatively more frequent at/near a given site of sequences of a given dataset than elsewhere in these sequences. Only one dataset is considered in this case.

- The second type is the *site-class-characteristic distinguishing sequence patterns*, defined as sequence patterns which are relatively more frequent at/near a given site of sequences of a target dataset than at/near the same site in sequences of an opposing dataset. In this case, two datasets are considered, together with a given site. For example, one may wish to find sequence patterns to distinguish the sequences around the transcription start site of one species against the sequences around the transcription start site of another species.
- The third type is the *class-characteristic distinguishing sequence pattern*, defined as sequence patterns which are relatively more frequent in sequences of a target dataset than in sequences in an opposing dataset. In this case, two datasets are considered, without a fixed site.
- The fourth type is the *surprising sequence patterns*, defined as sequence patterns whose occurrence in sequences of a given dataset is unexpected. In this case, there is just one dataset without a fixed site. This type of patterns are related to, but different from, the site-characteristic distinguishing sequence patterns in that (1) some sequences may not contain surprising sequence patterns, and (2) some sequences may contain multiple surprising sequence patterns. Surprising sequence patterns can indicate a rare mutation in DNA sequences of a species, or a surprising purchase behavior. A surprising sequence pattern can be considered significant if its occurrence frequency exceeds the prior expected frequency by a large margin.

Table 6.1. Four Distinguishing Sequence Pattern (DSP) Types

Pattern Type	Number of Datasets	Has-Site
site-characteristic DSPs	One	Yes
site-class-characteristic DSPs	Two	Yes
class-characteristic DSPs	Two	No
surprising sequence patterns	One	No

Thresholds are needed to formalize the meaning of "relatively more frequent". For example, for the site-characteristic distinguishing sequence patterns, one may require that a pattern be at least 10 times as frequent at the site as elsewhere. For the class-characteristic distinguishing sequence patterns, one may require that the patterns occur in some sequence in the target dataset but never occur in sequences in the opposing dataset. Some minimality condition may be imposed to reduce the number of patterns and to eliminate patterns which are too similar to others.

Sometimes the domain experts would first identify the sites and the datasets before data mining is performed. At other times, the sites and the datasets may need to be identified, together with the distinguishing patterns, from a universal set of sequences. Sometimes one is given the target dataset only, and may use the "background dataset" as the opposing dataset.

Some studies in the literature consider rare event patterns around some given site. Such rare event patterns can be classified as site-characteristic distinguishing sequence patterns.

Examples of distinguishing pattern types include protein domain motifs, transcription binding site motif, transcription binding site HMM, transcription binding site profile; these pattern types apply to one or two datasets with a given site. The mining of these pattern types was considered already in Chapter 4. Examples of class-distinguishing sequence patterns include distinguishing sequence patterns with gaps. Example applications of rare event patterns include the detection of fraud and alarms. Below we discuss some algorithms for the mining of these patterns.

6.2 Class-Characteristics Distinguishing Sequence Patterns

The discussion in this section is restricted to sequences where each element is a single item. Such sequences are able to capture some of the most important and popular sequences, such as DNA, proteins, documents and weblogs. This section is based on [55, 56].

6.2.1 Definitions and Terminology

Definition 6.2.1 ***(Subsequence Occurrence)*** *Given a sequence* $S = s_1 ... s_n$ *and a subsequence* $S' = s'_1 ... s'_m$ *of* S*, a set of positions* $\{i_1, i_2, ..., i_m\}$ *is called an occurrence of* S' *in* S *if* $1 \leqslant i_1 < ... < i_m \leqslant n$ *and* $s'_k = s_{i_k}$ *for each* $1 \leqslant k \leqslant m$*.* ∎

Example 6.2.1 *For the sequence* $S = ACACBCB$ *and subsequence* $S' = AB$*, there are* 4 *occurrences of* S' *in* S*:* $\{1, 5\}$*,* $\{1, 7\}$*,* $\{3, 5\}$ *and* $\{3, 7\}$*.* ∎

Gap constraints are now defined to restrict the allowed distance between items of subsequences in sequences.

Definition 6.2.2 ***(Gap constraint and satisfaction)*** *A (maximum) gap constraint is specified by a positive integer* g*. Given a sequence* $S = s_1 s_2 ... s_n$ *and an occurrence* $o_s = \{i_1, i_2, ..., i_m\}$ *of a subsequence* S' *in* S*, if* $i_{k+1} - i_k \leqslant g + 1$ *for all* $1 \leqslant k < m$*, then we say the occurrence* o_s *fulfills the* g*-gap constraint. Otherwise we say* o_s *fails the* g*-gap constraint. If there is at least one occurrence of a subsequence* S' *fulfilling the* g*-gap constraint, we say* S' *fulfills the* g*-gap constraint. Otherwise* S' *fails the* g*-gap constraint.* ∎

For Example 6.2.1, only the occurrence $\{3, 5\}$ fulfills the 1-gap constraint. Thus, the subsequence S' fulfills the 1-gap constraint since at least one of its occurrences does. No occurrence of S' fulfills the 0-gap constraint and so S' fails the 0-gap constraint.

Given a set of sequences D, a sequence pattern p and a gap constraint g, the count of p in D with g-gap constraint, denoted as $count_D(p, g)$, is the number of sequences in D in which p appears as a subsequence fulfilling the g-gap constraint. The (relative) support of p in D with g-gap constraint is defined as $supp_D(p, g) = \frac{count_D(p,g)}{|D|}$. Given a positive threshold δ, if $supp_D(p, g) \geqslant \delta$, we say p is frequent in D with g-gap constraint. Otherwise p is infrequent.

Definition 6.2.3 ***(g-MDS and the g-MDS mining problem)*** *Given two classes pos (the positive) and neg (the negative) of sequences, two support thresholds δ and α, and a maximum gap*[1] *g, a pattern p is called a Minimal Distinguishing Subsequence with g-gap constraint (g-MDS for short), if and only if the following conditions are true:*

Frequency condition: *$supp_{pos}(p, g) \geqslant \delta$;*
Infrequency condition: *$supp_{neg}(p, g) \leqslant \alpha$;*
Minimality condition: *There is no subsequence of p satisfying both the frequency condition and the infrequency condition.*

Given pos, neg, δ, α and g, the g-MDS mining problem is to find all the g-MDSs. ■

The minimality condition is important, since it both reduces output size and improves performance, as well as making patterns shorter (more succinct). This is especially important for datasets with long sequences, where the number of patterns output may be huge.

Table 6.2. Two sequence data classes

Sequence ID	Sequence	Class label
1	*CBAB*	*pos*
2	*AACCB*	*pos*
3	*BBAAC*	*pos*
4	*BCAB*	*neg*
5	*ABACB*	*neg*

Example 6.2.2 *Given the two sets (classes) of sequences shown in Table 6.2, suppose $\delta = 1/3$, $\alpha = 0$, and $g = 1$. The 1-MDSs are $\{BB, CC, BAA, CBA\}$. Notice that BB is a subsequence of all the negative sequences, if no gap constraint is used. However all the occurrences of BB in the negative sequences*

[1] We examine incorporation of a minimum gap constraint in Section 6.2.3. Also, in *ConSGapMiner*, the gap constraints for *pos* and *neg* do not necessarily have to be the same.

fail the 1*-gap constraint, so* BB *becomes a distinguishing subsequence when* $g = 1$*. Observe that every super sequence of a* 1*-MDS fulfilling the* 1*-gap constraint and support threshold is also distinguishing. However, these are excluded from the MDS set, since they are non-minimal and contain redundant information.* ∎

There are many situations where MDSs are useful, such as the comparison of proteins, design of microarrays (concerning the selection of DNA fragments), characterization of text and the building of classification models. The following two specific examples highlight the idea.

Example 6.2.3 *When comparing the two protein families zf-C2H2 and zf-CCHC, it was discovered that a protein section* $CLHH$ *appears as a subsequence* 141 *times among a total of* 196 *protein sequences in zf-C2H2, but it never appears among the* 208 *sequences in zf-CCHC. This subsequence represents a very strong contrast feature. From a classification perspective, an unknown protein sequence containing* $CLHH$ *as a subsequence seems unlikely to be a member of the zf-CCHC family.* ∎

The potential usefulness of distinguishing sequence patterns for protein datasets is highlighted by work in [95], where it is observed that biologists are very interested in identifying the most significant subsequences that discriminate between outer membrane proteins and non outer membrane proteins. Furthermore, the higher dimensional structure of proteins makes allowing gaps in a subsequence particularly important. Elements which have a gap between them in the sequence, may in fact be spatially very close in the 3-dimensional protein.

Example 6.2.4 *Comparing the first and last books from the Bible, it was found that the subsequences "having horns", "faces worship", "stones price" and "ornaments price" appear multiple times in sentences in the Book of Revelation, but never in the Book of Genesis. (The gap between the two words of each pair is* $\leqslant 6$ *non trivial words.) Such pairs might be seen as a fingerprint associated with the Book of Revelation and may be of interest to Biblical scholars.* ∎

6.2.2 The *ConSGapMiner* Algorithm

We now consider the *ConSGapMiner* algorithm, for solving the g-MDS mining problem. It has the following three main subroutines:

i) tree-based depth-first search framework to find a set of distinguishing sequences (containing all minimal distinguishing sequences),
ii) bitset based support and gap calculation, and
iii) post processing (minimization).

The first routine computes some set of distinguishing sequence patterns which contains all minimal distinguishing sequences. We call such a set of distinguishing sequence patterns a g-SMDS set, as defined below.

Definition 6.2.4 *(g-SMDS set) A* **Semi-Minimal Distinguishing Subsequence set** *with g maximum gap constraint, g-SMDS set for short, is a super set of the g-MDS set, such that elements in the g-SMDS set but not in the g-MDS set are sequence patterns that satisfy the frequency and infrequency conditions, but not necessarily the minimality condition.* ∎

A g-SMDS set may also contain some non-minimal distinguishing sequences, which will be removed in the minimization process in a batch manner. This choice was made since performing minimization whenever a new distinguishing sequence is generated is more expensive than batch-based minimization.

We now discuss each of the three routines in turn.

SMDS set generation

ConSGapMiner performs a depth-first search in a lexicographic sequence tree. In this tree, each node contains a sequence S, a value for $count_{pos}(S, g)$ and a value for $count_{neg}(S, g)$. Each node is the max-prefix[2] of each of its children. During the depth-first search, we extend the current node by a single item from the alphabet, according to a certain lexicographic order. For each newly-generated node v, we calculate the supports of v's associated sequence from *pos* and from *neg*. Part of the lexicographic tree for mining the data of Table 6.2 is given in Figure 6.1. Observe that the branches of the lexicographic tree terminate at nodes whose $count_{pos} = 0$.

Two basic pruning strategies can be used to reduce the size of the search space of the tree. These will be applied in the candidate generation process.

Non-Minimal Distinguishing Pruning: This strategy is based on the fact that any supersequence of a distinguishing sequence cannot be a minimal one. Suppose we encounter a node representing sequence S, where c is the last item in S, $supp_{pos}(S, g) \geqslant \delta$, and $supp_{neg}(S, g) \leqslant \alpha$. Then i) we never need to extend S and ii) we never need to extend any of the sibling nodes of S by the item c. Such an extension would lead to a supersequence of S which cannot be an MDS.
For Figure 6.1, since $supp_{pos}(AACC) > 0$ and $supp_{neg}(AACC) = 0$, $AACC$ must be distinguishing. Moreover, we know in the subtree of its sibling $AACB$, $supp_{neg}(AACBC)$ must be 0, too. So $AACBC$ can't be an MDS.

---

[2] The *max-prefix* of a sequence $S = s_1...s_n$ is $s_1...s_{n-1}$, formed by removing the last item in S. For example ABC is the max-prefix of $ABCD$ but AB isn't.

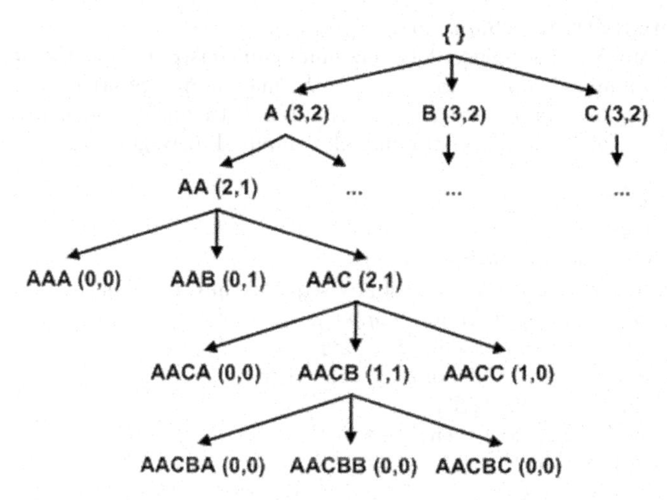

Fig. 6.1. Part of the lexicographic tree for mining Table 6.2

Max-Prefix Infrequency Pruning: Whenever a candidate is not frequent in *pos*, none of its descendants in the tree can be frequent. Thus, whenever we come across a sequence S at a node satisfying $supp_{pos}(S, g) < \delta$, we do not need to extend this node any further. For example, in Figure 6.1, it is not necessary to extend AAB (which has support zero in *pos*), since no frequent sequence can be found in its subtree.

It is worth noting that this technique does not generalize to full a-priori like pruning – "if a subsequence is infrequent in *pos*, then no supersequence of it can be frequent". Such a statement is not true, because the gap constraint is not class preserved [130]. This means that an infrequent sequence's supersequence is not necessarily infrequent; this consequently increases the difficulty of the MDS mining problem. Indeed, extending an infrequent subsequence by appending will not lead to a frequent sequence, but extensions by inserting items in the middle of the subsequence may lead to a frequent subsequence.

An example situation is given next. For Figure 6.1, suppose $\delta = 1/3$ and $g = 1$. Then AAB is not a frequent pattern because $count_{pos}(AAB, 1) = 0$. But looking at AAB's sibling, the subtree rooted at AAC, we see that $count_{pos}(AACB, 1) = 1$. So a supersequence $AACB$ is frequent, but its subsequence AAB is infrequent.

The SMDS set generation algorithm is given in Figure 6.2. The algorithm is called at the root of the search tree by Candidate_Gen($\{\}, g, I, \delta, \alpha$), with SM initialized to be the emptyset.

Algorithm: SMDS_Gen(S,g,I,δ,α);
Assumption: S is a sequence, g is maximum gap constraint, I is the alphabet,
δ is the minimal support for *pos*, α is the maximum support for *neg*;
CDS is a local variable storing the children distinguishing sequences of S;
SM is a global variable containing all computed distinguishing subsequences;
Method:
1: initialize CDS to {};
2: for each $x \in I$ do
3: let $S' = S.x$ (appending x to S);
4: if S' is not a supersequence of any sequence in CDS then
5: $supp_{pos}$=Support_Count(S',g,pos);
6: $supp_{neg}$=Support_Count(S',g,neg);
7: if ($supp_{pos} \geqslant \delta$ AND $supp_{neg} \leqslant \alpha$) then
8: $CDS = CDS \cup \{S'\}$;
9: elsif ($supp_{pos} \geqslant \delta$) then
10: SMDS_Gen(S',g,I,δ,α);
11: $SM = SM \cup CDS$;

Fig. 6.2. The *SMDS_Gen* routine

Support Calculation and Gap Checking

For each newly-generated candidate S, $count_{pos}(S, g)$ and $count_{neg}(S, g)$ must be computed. The main challenge comes in checking satisfaction of the gap constraint. A candidate can occur many times within a single sequence. A straightforward idea for gap checking would be to record the occurrences of each candidate in a separate list. When extending the candidate, a scan of the list determines whether or not the extension is legal, by checking whether the gap between the end position and the item being appended is smaller than the (maximum) gap constraint value for each occurrence. This idea becomes ineffective in situations with small alphabet size and small support threshold and many long sequences needing to be checked, since the occurrence list becomes unmanageably large. Instead, a more efficient method for gap checking can be used, based on a bitset representation of subsequences and the use of boolean operations. This technique is described next.

Definition 6.2.5 ***(Bitset)*** *A bitset is a sequence of binary bits. An n-bitset X contains n binary bits, and $X[i]$ refers to the i-th bit of X.* ■

A bitset can be used to describe how a sequence occurs within another sequence. Suppose we have two sequences $S = s_1...s_n$ and $S' = s'_1...s'_m$, where $m \leqslant n$. The occurrence(s) of S' in S can be represented by an n-bitset. This n-bitset BS is defined as follows: If there exists a supersequence of S' of the form $s_1...s_i$ such that $s_i = s'_m$ (the last item of S'), then $BS_{[i]}$ is set to 1; otherwise it is set to 0. For example, if $S = BACACBCCB$, the 9-bitset representing $S' = AB$ is 000001001. This indicates how the subsequence AB can occur in $BACACBCCB$, with a '1' being turned on in each final position where the subsequence AB could be embedded. If S' is not a subsequence of S, then the bitset representing the occurrences of S' consists of all zeros.

For the special case where $S' = s$ is a single item, $BS_{[i]}$ is set to 1 if $s_i = s$. For $S = BACACBCCB$, the 9-bitset representing $S' = C$ is 001010110.

It will be necessary to compare a given subsequence against multiple other sequences. In this case, the subsequence will be associated with an array of bitsets, where the k-th bitset describes the occurrences of S' in the k-th sequence.

Initial Bitset Construction: Before mining begins, it is necessary to construct the bitsets that describe how each item of the alphabet occurs in each sequence from the *pos* and *neg* datasets. So, each item x has associated with it an array of $|pos| + |neg|$ bitsets; the number of bitsets in x's array which contain one or more 1's, is equal to $count(x, g)$.

For the data in Table 6.2, the bitset array for A contains 5 bitsets, namely $[0010, 11000, 00110, 0010, 10100]$. Moreover, $count_{pos}(A, g) = 3$ and $count_{neg}(A, g) = 2$.

Bitset Checking: Each candidate sequence S in the lexicographic tree has a bitset array associated with it, which describes how S can occur in each of the $|pos| + |neg|$ sequences. This bitset array can be directly used to compute $count_{pos}(S, g)$ and $count_{neg}(S, g)$ (i.e. $count_{pos}(S, g)$ is just the number of bitsets in the array not equal to zero, that describe positive sequences). During mining, we extend a sequence S (at a node) to get a new candidate S', by appending some item x. Before computing $count_{pos}(S', g)$ and $count_{neg}(S', g)$, we first need to compute the bitset array for S'. The bitset array for S' is calculated using the bitset array for S and the bitset array for item x, and is done in two stages.

Stage 1: Using the bitset array for S, we generate another array of corresponding mask bitsets. Each mask bitset captures all the valid extensions of S, with respect to the gap constraint, for a particular sequence in $pos \cup neg$. Suppose the maximum gap is g. For a given bitset b in the bitset array of S, we perform $g+1$ times of right shift by distance 1, with 0s filling the leftmost bits. This results in $g+1$ intermediate bitsets, one for each stage of the shift. By ORing together all the intermediate bitsets, we obtain the final mask bitset m derived from b. The mask bitset array for S consists of all such mask bitsets.

Example 6.2.5 *Taking the last bitset* 10100 *in the previous example and setting* $g = 1$, *the process is:*

$$\begin{array}{rl} 10100 >> & 01010 \\ 01010 >> & 00101 \\ \hline OR & 01111 \end{array}$$

01111 *is the mask bitset derived from bitset* 10100. ■

Intuitively, a mask bitset m generated from a bitset b, closes all 1s in b (by setting them to 0) and opens the following $g+1$ bits (by setting them to 1). In this way, m can accept only 1s within a $g+1$ distance from the 1s in b.

Stage 2: We use the mask bitset array for S and the bitset array for item x, to calculate the bitset array for S' which is the result of appending x to S. Consider a sequence X in $pos \cup neg$ and suppose the mask bitset describing it is m and the bitset for item x is t. The bitset describing the occurrence of S' in X, is equal to m AND t. If the bitset of the new candidate S' does not contain any 1, we can conclude that this candidate is not a subsequence of X with g-gap constraint.

Example 6.2.6 *ANDing* 01111 *(the mask bitset for sequence A) from the last example with C's bitset* 00010, *gives us* AC*'s bitset* 00010.

Taking the last sequence in Table 6.2, ABACB, B's 5*-bitset is* 01001 *and its mask* 5*-bitset is:*

$$\begin{array}{ll} 01001 >> 00100 \\ 00100 >> 00010 \\ \hline \textit{OR} \quad\quad 00110 \end{array}$$

So BB's bitset is: 00110 *AND* 01001 = 00000. *This means BB is not a subsequence of ABACB with* 1*-gap constraint.* ■

Example 6.2.7 *Figure* 6.3 *shows the process of getting the bitset array for BB from that for B. The two tables on the two sides of the arrow ⇒ show how the masks for B are obtained from the bitset array for B. The & operation is taken on bitset array and the Masks set, yielding the bitset array for BB. From the figure we can see* $count_{pos}(BB, 1) = 2$ *and* $count_{neg}(BB, 1) = 0$. ■

Bitsets of B		Masks of B		Bitsets of B		Bitsets of BB
0101		0011		0101		0001
00001	⇒	00000	&	00001	=	00000
11000		01110		11000		01000
1001		0110		1001		0000
01001		00110		01001		00000

Fig. 6.3. The generation of BB's bitset array.

The task of computing bitset arrays can be done very efficiently. Modern computer architectures have very fast implementations of shift operations and logical operations. Since the maximum gaps are usually small (e.g. less than 20), the total number of right shifts and logical operations needed is not too large. Consequently, calculating $supp_{pos}(S, g)$ and $supp_{neg}(S, g)$ can be done extremely quickly. The algorithm for support counting is given in Figure 6.4.

Minimization

We have already seen how non-minimal distinguishing pruning eliminates non-minimal candidates during tree expansion. However, the pattern set returned

Algorithm: Support_Count(S', g, D);
Assumption: g is the maximum gap, S is the max-prefix of S', the bitset array $BARRAY_S$ for S is available, the bitset array $BARRAY_x$ for the final element x of S' is available, D is the dataset;
Output: $supp_D(S', g)$ and $BARRAY_{S'}$ (the bitset array for S');
Method:
1: generate the mask bitsets $MaskS$ from $BARRAY_S$ for g (stage 1 above);
2: do bitwise AND of $MaskS$ and $BARRAY_x$ to get $BARRAY_{S'}$ (stage 2 above);
3: let *count* be the number of bitsets in $BARRAY_{S'}$ which contain 1;
4: return $supp_D(S', g) = count/|D|$ and $BARRAY_{S'}$;

Fig. 6.4. Support_Count(S',g,D): calculate $supp_D(S', g)$

by Algorithm 6.2 is only semi-minimal, i.e. an SMDS set. For example, in Figure 6.1, we will get ACC, which is a supersequence of the distinguishing sequence CC. Thus, in order to get the g-MDS set, a post-processing minimization step is needed.

A naïve idea for removing non-minimal sequences, is to check each against all the others, removing it if it is a supersequence of at least one other. For n sequences, this leads to an $O(n^2)$ algorithm, which is expensive if n is large.

Two ideas can be used to make it more efficient. Firstly, observe that it is not necessary to check if a sequence is a supersequence of any longer sequence. To take advantage of this, we cluster the sequence patterns according to their length, when they are output during mining.

Secondly, we use a prefix tree for carrying out minimization. Sequences are inserted into the tree in ascending order of length. Each sequence S to be inserted into the tree is compared against the sequences already there. This is easily done by stepping through each prefix of S, at each stage identifying the nodes of the tree which are subsequences of the prefix so far. The process terminates when a leaf node or the end of S is reached. If a subsequence of S in the tree is found, then S is discarded. Otherwise, S must be minimal and it is inserted.

Compared to the naive $O(n^2)$ method, using a prefix tree can help avoid some duplicate comparisons, particularly for situations where there is substantial similarity between the sequential patterns, since each sequential pattern prefix is only stored once. For example, consider two shorter patterns $P_1 = ABCC$, $P_2 = ABCF$ and a longer pattern $P_3 = ABCDE$. To check whether P_3 is minimal by using the naive way, we compare P_3 with P_1 itemwise for 5 comparisons and with P_2 itemwise for 5 comparisons to conclude that P_3 is minimal. By using the prefix tree, ABC is built once and compared once, which takes itemwise 3 comparisons and then another 2 comparisons to check the other two items D and E. Finally we know that P_3 is minimal, because no leaf is found. This takes itemwise 5 comparisons total, rather than 10 comparisons using the naive way.

The complete algorithm of *ConSGapMiner* is provided in Figure 6.5.

Algorithm: ConSGapMiner(pos,neg,g,δ,α)
Assumption: I is the alphabet list, g is the maximum gap constraint,
δ is the minimal support in pos, α is the maximal support in neg,
a global set SMDS is used to contain the patterns generated by SMDS_Gen;
Output: g-MDS set MDS;
Method:
1: $SMDS \leftarrow \{\}$;
2: set S to the empty sequence;
3: SMDS_Gen(S,g,I,δ,α);
4: let MDS be the result of minimizing $SMDS$ as described above;
5: return MDS;

Fig. 6.5. The ConSGapMiner algorithm

6.2.3 Extending ConSGapMiner: Minimum Gap Constraints

This subsection and the next discuss several extensions to the ConSGapMiner algorithm. These involve handling minimum gaps (this subsection) and performing more complex types of minimization (the next subsection).

The definition of minimum gap constraint is essentially the dual of the maximum gap constraint, obtained by replacing $\leqslant$ with $\geqslant$.

Definition 6.2.6 ***(Minimum Gap Constraint)*** *A minimum gap constraint is specified by a positive integer q. Given a sequence $S = s_1...s_n$ and an occurrence $o_s = \{i_1, i_2, ..., i_m\}$ of a subsequence S', if $i_{k+1} - i_k \geqslant q + 1$ for each $k \in \{1, ..., m-1\}$, then the occurrence o_s is said to fulfill the q-minimum gap constraint. Otherwise we say o_s fails the constraint. If there is at least one occurrence of a subsequence S' fulfilling the q-minimum gap constraint, we say S' fulfills the q-minimum gap constraint. Otherwise S' fails the constraint.* ■

Example 6.2.8 *Consider a sequence $S = ACEBE$. If a maximum gap $g = 3$ is given, then AE has 2 subsequence occurrences in S fulfilling the constraint, namely $\{1, 3\}$ and $\{1, 5\}$. If a minimum gap $q = 2$ is given, then AE has one occurrence $\{1, 5\}$ in S fulfilling the constraint. So AE is a subsequence of S fulfilling both the maximum and the minimum constraints. Considering only the above maximum gap constraint, AC is a subsequence of S fulfilling the constraint; but in conjunction with the above minimum gap constraint, AC is not a subsequence of S fulfilling the constraints.* ■

Minimum gaps can be useful for applications where items of a sequence pattern need to be at least certain distance apart from one another in sequences. For example, in scenarios where the items in the sequence represent values being sampled over time, such as a waveform, items that are too close to each other may represent information that is overly similar. Minimum gaps may then be specified to help remove potential redundancy in the discovered patterns. An interesting special case is when the value of the minimum gap is specified to equal the value for the maximum gap. This will result in patterns whose items are distributed in equal distance in the original dataset.

Suppose we have a set of sequences D, a sequential pattern p and two gap constraints g and q. The count of p in D with g as the maximum gap constraint and q as the minimum gap constraint, denoted as $count_D(p, g, q)$, is the number of sequences in D in which p appears as a subsequence fulfilling both the g-gap constraint and the q-minimum gap constraint. The (relative) support of p in D with g-gap constraint and q-minimum gap constraint is defined as $supp_D(p, g, q) = \frac{count_D(p,g,q)}{|D|}$.

We now redefine the mining problem to include both the minimum and maximum gap constraints:

Definition 6.2.7 ***(Extended MDS mining problem)*** *Given two classes of sequences pos (the positive) and neg (the negative), two support thresholds δ and α, a maximum gap g and a minimum gap q, a pattern p is called a Minimal Distinguishing Subsequence with (g, q)-gap constraint ((g, q)-MDS for short), if and only if the following conditions are true:*

- ***Frequency condition:*** $supp_{pos}(p, g, q) \geqslant \delta$;
- ***Infrequency condition:*** $supp_{neg}(p, g, q) \leqslant \alpha$;
- ***Minimality condition:*** *There is no subsequence of p satisfying both the frequency condition and the infrequency condition.*

Given pos, neg, δ, α, g and q, the (g, q)-MDS mining problem is to find all the (g, q)-MDSs. ■

It is easy to extend the bit operations used in *ConSGapMiner* to handle this minimum gap constraint. The only part requiring modification is the construction of the mask bitset. Recall that for a maximum gap constraint g, to construct the mask bitset, we perform $g + 1$ times of right shift by distance 1 and OR the $g + 1$ intermediate bitsets together. The resulting bitset is the mask of the given bitset. When a minimum gap constraint q is added, we initially right shift the given bitset q times and discard the intermediate bitsets. Then, we perform another $g + 1 - q$ right shift operations and OR the $g + 1 - q$ intermediate bitsets to obtain the mask bitset.

Example 6.2.9 *Consider the last sequence in Table 6.2 as an example. We know that* B*'s bitset is* 00111 *w.r.t.* $g = 2$. *So* BC*'s bitset should be:* 00111 *AND* 00010=00010. *For* $q = 2$, *we discard the first 2 intermediate bitsets which are* 00100 *and* 00010. *Then we right shift* $g + 1 - q = 1$ *times and the mask bitset is* 00001. *From this bitset we can see that the adjacent two positions, which are* 3 *and* 4, *are closed and the third position* 5 *is open. This mask bitset expresses the minimum gap constraint. By ANDing this mask bitset with the single item C's bitset* 00010, *we get* 00000; *so* BC *isn't a subsequence of* ABACB, *because it fails the minimum gap constraint* $q = 2$. ■

6.2.4 Extending ConSGapMiner: Coverage and Prefix-Based Pattern Minimization

In order to discover patterns which satisfy the minimality condition from Definitions 6.2.3 and 6.2.7, we have described strategies for pattern minimization which aim to determine whether a pattern (or candidate) is a supersequence of some other pattern (or candidate). For some situations, pursuing this kind strategy may be too aggressive and useful patterns may be eliminated, since the gap constraint is not class preserved. Consider the following example:

Example 6.2.10 *Suppose* pos={ACCBD, ACDBD, ABD} *and two distinguishing patterns* ACBD *and* ABD *have been found using* $\delta = 1/2$, $g = 1$, $q = 0$. *We observe that* $count_{pos}(ACBD, 1, 0) = 2$ *and* $count_{pos}(ABD, 1, 0) = 1$. *If we remove* ACBD *because it is a super-sequence of* ABD, *we will lose a pattern which has higher frequency than its subpattern and thus is arguably a more important feature.* ■

To address this problem, two alternate minimization techniques can be used. The first is based on comparisons using coverage in *pos*, the second is based on prefix comparisons. Both are less aggressive and remove fewer patterns than the minimization described above (which will be called *basic minimization* from now on).

In essence, we need to use stricter conditions for minimality than that used in Definitions 6.2.3 and 6.2.7. Given two MDSs, p_1 and p_2 and a reference dataset D, we wish to remove p_2 due to p_1, if p_1 occurs in every sequence from D that p_2 does ($p_2 \Rightarrow_D p_1$). If p_1 occurs in every sequence from *pos* that p_2 does, then it is guaranteed that ($p_2 \Rightarrow_D p_1$) only if $D = pos$. This is *coverage based minimization.* If p_1 is prefix of p_2, then it is guaranteed that $p_2 \Rightarrow_D p_1$, with respect to any D. This is *prefix based minimization.*

Handling Coverage Based Minimization

To describe the implementation of coverage based minimization, we first begin by formally defining the notion of coverage.

Definition 6.2.8 ***(Coverage Set)*** *Given a sequence p, gap parameters g and q and the datasets* pos *and* neg, *the coverage set of p is equal to the set of sequences from pos in which p appears as a subsequence fulfilling both the g-maximum gap constraint and the q-minimum gap constraint. The coverage set can be represented by a bitset, containing as many bits as there are sequences in pos. Bit i is turned on if p appears as a subsequence in sequence i from pos. Otherwise it is set to zero.* ■

When performing coverage based minimization, a sequence p_1 can be eliminated by a sequence p_2 iff:

1. p_2 is a subsequence of p_1 and

2. the coverage set of p_1 is a subset of the coverage set of p_2.

To adjust the basic *ConSGapMiner* algorithm from Section 6.2.2 we have to change two techniques: the first is the non-minimal distinguishing pruning step and the other is the post-processing minimization.

For each candidate in the lexicographic tree shown in Figure 6.1, a coverage bitset is attached. This coverage bitset contains as many bits as the total of sequences in *pos*. A newly-generated candidate's coverage bitset can be set by its bitset array. If the i-th bitset in the array contains at least one 1, this candidate's $count_{pos}$ is increased and the i-th bit in the coverage bitset is set to 1. The rule for the pruning is then changed to:

Non-Minimal Distinguishing Pruning (adjusted): Suppose we encounter a node representing sequence S, where c is the last item in S and $supp_{pos}(S, g) \geqslant \delta$ and $supp_{neg}(S, g) \leqslant \alpha$. Then i) we never need to extend S and ii) for any of the sibling nodes S', we AND the coverage bitset for S' with the coverage bitset for S and then XOR the resulting bitset with the coverage bitset for S'. If the resulting bitset does not contain any 1, then we never need to extend S' by the item c. If the resulting bitset contains at least one 1, we must extend S' by the item c.

Boolean operations can be used to test whether S's coverage is a superset of S''s. If this is the case, then no candidate in the subtree of S' has a coverage which is not a subset of S's. In this situation, any extension of the nodes in the subtree of S', with item c can give a super-sequence of S with a subset of S's coverage, which means it is not minimal. If S' has a non-subset coverage of S, then there may be an extension with item c, which gives a distinguishing subsequence with a non-subset coverage of S's and this may be minimal, so we need to generate and keep it.

For the post-processing minimization, we still order the patterns in descending order of their lengths. For each pattern we still keep a coverage bitset with the same meaning as described above. For each pattern p, we find all the patterns which have shorter lengths and with a coverage that is a superset of that of p. If such a pattern is found, the standard way of checking whether this pattern is a subsequence of p or not is performed. If it is a subset, p is eliminated. If no pattern can be found to eliminate p, then we retain it in the MDS set.

Handling Prefix Based Minimization

Performing prefix based minimization is substantially simpler than performing coverage based minimization. Two modifications are needed to the basic ConsSGapMiner algorithm. First of all, the non-minimal distinguishing pruning step is adjusted to the following:

Non-Minimal Distinguishing Pruning (adjusted): Suppose we encounter a node representing sequence S, where c is the last item in S and

$supp_{pos}(S, g) \geqslant \delta$ and $supp_{neg}(S, g) \leqslant \alpha$. Then we do not need to extend S.

Secondly, no post processing minimization step is needed, since all distinguishing sequences produced are guaranteed to be prefix minimal.

6.3 Surprising Sequence Patterns

Roughly speaking, a surprising sequence pattern is one whose (frequency of) occurrence deviates greatly from the prior expectation of such occurrence.

Definition 6.3.1 *Given a dataset D of sequences, a surprisingness measure θ on sequence patterns, and a surprisingness threshold $minSurp$, a sequence pattern S is called a* surprising sequence pattern *if $\theta(S) \geqslant minSurp$ holds.* ∎

Essentially, the surprisingness $\theta(S)$ of a sequence pattern S can be defined as the difference of the actual frequency and the expected frequency of S. A pattern S's actual frequency of occurrence can be surprising if the actual frequency is much higher or much lower than the expected frequency. A pattern can be considered surprising with respect to a dataset, or with respect to a sequence, or with respect to a window of a sequence, depending on how the actual frequency is calculated. In the first case, the actual frequency is calculated from the entire dataset. In the second case, the actual frequency is calculated from the given sequence. In the third case, the actual frequency is calculated from a window of the given sequence. The first case can be useful for identifying some unexpected behavior of a class of sequences (e.g. the sequences of a species). The second and third cases are useful for predicting surprisingness of (windows of) sequences.

Several issues need to be addressed for surprisingness analysis:

- One issue is how to define the frequency (actual or expected) of sequence patterns. Possibilities include (1) per-sequence based definition where all occurrences of a pattern in an entire sequence are considered, and (2) per-window based definition where all occurrences of a pattern in a window of a given sequence are considered. Per sequence/window based frequencies can be used to determine frequencies in a whole dataset if desired.
- Another issue is how to estimate the expected frequency (or probability) of a sequence pattern.
- A third issue is how to choose the $minSurp$ threshold. This issue is important for avoiding having too many false positive "surprising sequence patterns", especially for alarm/fraud detection applications.

Below we discuss how to estimate the expected frequency for the window based definition of frequency. This essentially follows the approach of [41]. (Reference [129] deals with the same issue for periodic sequence patterns.) A window size is a positive integer w.

Definition 6.3.2 *Let $S = s_1...s_m$ be a sequence pattern and let $T = t_1...t_n$ be a sequence. Then the w-window frequency of S in T is defined as the total number, denoted as $C(n, w, m)$, of occurrences (matches) $\{i_1, ..., i_m\}$ of S in T such that $i_1 < ... < i_m$, $s_j = t_{i_j}$ for all $1 \leqslant j \leqslant m$, and $i_m - i_1 \leqslant w$ (i.e. the distance between the first matching position and the last matching position of the pattern in the sequence is at most w).* ∎

To estimate the expected window based frequency of S in D, it is convenient to think of D as having just one sequence. This can be done by concatenating all sequences of D into one sequence. Moreover, it is convenient to assume that the concatenated sequence is generated by a memoryless (Bernoulli) source. It is worth remembering that the expected frequency is just an estimation, although one wants to get accurate estimates.

For each item a, let $prob(a)$ be the estimated probability of a in the dataset D. For example, $prob(a)$ can be estimated as

$$prob(a) = \frac{\text{number of occurrences of } a \text{ in } D}{\Sigma_{X \in D}|X|}.$$

Let $p(w, m)$ denote the probability that a window of size w contains at least one occurrence of the sequence pattern S of size m as a subsequence. Then the expected value of $C(n, w, m)$ can be estimated as

$$E(C(n, w, m)) = n \times p(w, m).$$

Reference [41] shows that, for sufficiently large w, $p(w, m)$ can be approximated by an expression in terms of $prob(a)$ of all items a in the alphabet. Hence $E(C(n, w, m))$ can be approximated by an expression in terms of $prob(a)$ of all items a in the alphabet as well.

Reference [41] also computes the variance of $C(n, w, m)$, and then shows that $C(n, w, m)$ is normally distributed. This allows us to set either an upper threshold $\tau_u(w, m)$ (for over-represented surprising patterns) or a lower threshold $\tau_\ell(w, m)$ (for under-represented surprising patterns). More precisely, for a given level β, the choice of $\tau_u(w, m)$ and $\tau_\ell(w, m)$ ensures that either $P(\frac{C(n,w,m)}{n} > \tau_u(w, m)) \leqslant \beta$ or $P(\frac{C(n,w,m)}{n} < \tau_\ell(w, m)) \leqslant \beta$. That is, if one observes more than $n\tau_u(w, m)$ occurrences (upper threshold) or fewer than $n\tau_\ell(w, m)$ occurrences (lower threshold) of windows with sequence pattern S, it is highly unlikely that such a number is generated by the memoryless source, which implies that S is surprising. The interested readers should consult [41] for details.

In addition to the memoryless model for approximating the data source, one can also use other models such as Markov models or HMM etc.

We end this section by discussing the related topic of "rare case mining" [120]. This applies to all types of data. Informally, a case corresponds to a region in the instance space that is meaningful with respect to the domain under study and a rare case is a case that covers a small region of the instance space and covers relatively few training examples. As a concrete example, with respect to the class bird, non-flying bird is a rare case since very few birds (e.g., ostriches) do not fly. Rare cases can also be rare associations. For example, mop and broom will be a rare association (i.e., case) in the context of supermarket sales [66]. For classification tasks the rare cases may manifest themselves as small disjuncts (namely rules that cover few training examples, where a rule is the set of conditions in a path from the root of a decision tree to a leaf of the tree). In the study of rarity, one needs to decide whether the rarity of a case should be determined with respect to some absolute threshold number of training examples (absolute rarity) or with respect to the relative frequency of occurrence in the underlying distribution of data (relative rarity).

7

Related Topics

In this chapter we discuss some other topics on or related to sequence data mining, including structured-data mining, partial periodic patterns, and bioinformatics. We also briefly discuss sequence alignment, which is needed for understanding certain materials in several previous chapters. Finally, we give some pointers to biological sequence databases and resources.

More general information on data mining can be found in [44, 105, 46]. Reference [40] covers general algorithms on sequences and trees, and reference [25] covers biological sequence analysis.

7.1 Structured-Data Mining

Sequence data mining can be viewed as a special subfield of data mining of structured data. By structured data we mean data where each data object is an explicitly structured composition of a set of data items. In addition to sequence data, other types of structured data include tree data, graph data, time series data, and text data. The explicit structures for the underlying data items in such structured data objects include (partial) orderings, temporal orderings, hierarchical structures, and network structures. It should be noted that our concept of structured is different from the concept of structured in databases: we mean that the lower level items in a high level object are somehow organized, whereas the concept of structured in databases means that the high-level objects are "typed" using the tuple and set constructs.

Graph data mining is useful for applications where general interactions among data items (objects) are of interest. For example, graph data mining has applications in the analysis of massive biological networks, chemical compounds and biological molecules, social networks, software and program structures (e.g. for debugging), and web structures, etc. Reference [119] is a survey on this field, although substantial progress has happened since the time of the survey. References [21, 124] are two representative papers in this area.

Tree data mining is useful for applications where hierarchical interactions among data items (objects) are of interest. For example, tree data mining has applications in XML data analysis, text data parsing and analysis. References [131, 18] are representative papers in this area.

Time series data mining has applications for situations where a huge number of time series (or data streams) are present. Here one would like to understand how different time series relate to each other, or how different parts of a time series relate to each other. When the time series consist of just numerical values over time, the main goal is to understand the trends. When the time series consist of sequences of events or composite objects over time, it is of interest to understand the interaction among the events and objects in addition to trends. Many diverse problems and approaches have been studied. References [9, 61, 52, 17] are representatives of papers in this area.

From the viewpoint of structured data, text data can be viewed as having a mixture of two structures, namely the sequencing structure at the sentence level and a hierarchical structure at the paragraph/section levels. Text data mining has a rich history and continues to be a fertile area for data mining research. Interested readers can use [48] as a start point to get more details.

Transactional data and relational data should not be viewed as structured. It is of little, if any, interest to mine structural relationship among the lower level items in such data.

7.2 Partial Periodic Pattern Mining

Finding periodic patterns from time series databases is an important data mining task with many applications, including the potential for predicting future events. Many methods have been developed for searching periodicity patterns in large data sets [70]. Periodicity can be full periodicity or partial periodicity. In full periodicity, every time point contributes (precisely or approximately) to the cyclic behavior of a time series. In contrast, in partial periodicity, some (not necessarily all) time points contribute (precisely or approximately) to the cyclic behavior of the time series. An example partial periodic pattern may state that Jim reads the New York Times from 7:00 to 7:30 every weekday morning but his activities at other times do not have much regularity. Thus, partial periodicity is a looser kind of periodicity than full periodicity, and it exists ubiquitously in the real world. The concept of partial periodic pattern was first introduced in [43, 42], and many results have been published since then.

Typically, the study of partial periodicity considers a single long event sequence associated with some real-world object. Often, each sequence position is associated with a time point. However, it should be noted that partial periodicity can also be applied to biological sequences, where one aims to discover partial periodicity over the sequence positions.

Many types of event sequences can be studied, leading to partial periodicity patterns with varying degree of structural complexity:

- The elements of sequences can be single items. A given sequence for partial periodicity study has the form $s_1...s_n$ where each s_i is a single item. Examples of such sequences include DNA sequences, and action histories for individual persons.
- The elements of sequences can be sets of items (or equivalently transactions). A given sequence for partial periodicity study has the form $s_1...s_n$ where each s_i is a set of items. Examples of such sequences include a customer shopping history, where each sequence position is associated with a set of items bought on one shopping trip, at a supermarket.
- The elements of sequences can be sets of transactions. A given sequence for partial periodicity study has the form $s_1...s_n$ where each s_i is a set of transactions. Examples of such sequences include the transaction history of a supermarket, where each sequence position is associated with the set of all customer transactions for some given day.

Of course, partial periodicity can also be studied for even more complex types of event histories (e.g. each sequence position is associated with a relation).

For simplicity, we now provide the formal definitions for the case where the elements of event sequences are sets of items, as given in [42]. For ease of discussion, we may refer to each sequence position as a time point.

Assume that a sequence of the form $S = s_1...s_n$ is given, where each s_i is a set of items. Let Σ be the underlying set of items that can occur in S. For example, S can be $a\{b,c,d\}cb\{b,c,e\}da\{b,c,f\}e$, for which Σ can be $\{a,b,c,d,e,f\}$.

We will also use the wildcard character $*$, which can match any single set of features at any given sequence position. We define a pattern $T = t_1...t_p$ as a non-empty sequence over $(2^{\Sigma} - \{\emptyset\}) \cup \{*\}$; we say p is the period of T. For example, $a\{b,c\}*$ is a pattern with period 3. For simplicity, we omit the parenthesis for singleton sets; for example, we write $\{a\}$ as a.

The frequency count of pattern T in sequence S is defined by

$$freq\_count(T,S) = |\{i \mid 0 \leqslant i < m, \text{ and } t_j \subseteq s_{i*p+j} \text{ for each } 1 \leqslant j \leqslant p\}|,$$

and the confidence of T in S is defined by

$$conf(T,S) = \frac{freq\_count(T,S)}{m},$$

where m is the maximum number of disjoint intervals of length p contained in S. (In formula, m is the positive integer such that $m*p \leqslant n < (m+1)*p$.)

The goal of partial periodic pattern mining is to mine the partial periodic patterns from S meeting a given minimum confidence threshold. When the period p is not given, one also needs to determine the periods for which there are partial periodic patterns from S meeting the given minimum confidence threshold.

Reference [43] introduced the partial periodicity mining problem for sequences over simple items, and gave a data cube based algorithm. Reference [42] provided the definition used above for generalized sequences. It also gave a mining algorithm, based on a so-called max-subpattern hit set property, which performs shared mining for a range of desired periods.

After these two papers, a number of papers have been published.

- Several papers focused on efficiency issues. Reference [27] considered the incremental mining of partial periodic patterns when a given sequence is extended, together with a "merge mining" partial periodic pattern mining algorithm. Reference [28] proposed a "convolution" based algorithm to mine partial periodic patterns with unknown periods in one pass. In this way, the periods for partial periodic patterns do not have to be user-specified, and the detection of the periods and the associated patterns can be done in one pass. Reference [29] gave another algorithm solving the same problems. Reference [14] also studied efficiency issues for partial periodic pattern mining.
- Several papers considered various extensions to partial periodicity. Reference [128] studied how to mine high-level partial periodic patterns, where a higher level pattern may consist of repetitions of lower level patterns. The paper overcame difficulties caused by the presence of noise for the discovery of high-level patterns. The paper used meta-patterns to capture these high level patterns. Reference [129] studied the mining of surprising periodic patterns. The paper used "information gain" to measure the degree of surprise. References [73, 127] studied the detection of asynchronous partial periodic patterns. Such patterns are useful for situations where the occurrences of a periodic pattern may be shifted due to disturbance (and so may not exactly match the period). Reference [106] studied the relaxed concept of periodicity, where a period may be stretched or shortened in a sequence.

Partial periodic pattern mining is related to the mining of cyclic association rules [82], the discovery of calendar-based temporal association rules [65], etc. It is also related to the analysis of time series (many text books, including [10] are available on this topic) and temporal knowledge discovery (see [93] for a survey on this topic).

7.3 Bioinformatics

Sequence data mining is closely related to bioinformatics and computational biology. Bioinformatics and computational biology are concerned with the development of computational tools for the management, analysis and understanding of biological data. Research in these directions are motivated to solve problems inspired from biological, genomic and medical practices.

In addition to sequence data, several other types of data have been collected from biological, genomic, and medical studies. Examples include the following: (a) Microarray gene chips can simultaneously profile the expression levels of thousands of genes in a single tissue sample. Such chips are a useful tool for understanding the interaction among genes, and the difference of such interactions under different disease and treatment conditions. (b) Tandem mass spectrometry is a process in which proteins are broken up and the numerous pieces (the peptides) are separated by mass. The result is a collection of tandem mass spectra, each of which is produced by a peptide and can act as a fingerprint for identifying the peptide. The data produced by tandem mass spectrometry can be used to understand the interaction among proteins in cells. (c) For many proteins, the three dimensional structure data are also available. We can study such data together with the protein sequence (linear structure).

In the previous chapters we touched on several types of bioinformatics problems such as gene start site identification problem, alternative splicing site identification problem, the motif finding and scoring problems, protein sequence family identification and characterization problems. There are many other bioinformatics problems. Some examples[1] are the following:

- Protein structure prediction problem: Given a protein amino acid sequence (a linear structure), determine its three-dimensional folded shape (a tertiary structure).
- Protein folding pathway prediction problem: Given a protein amino acid sequence and its three-dimensional folded structure, determine time-ordered sequence of folding events, called the folding pathway, that leads from the linear structure to the three-dimensional structure.
- Similar sequence search: Given a sequence S and a set of sequences $\mathcal{S}$, retrieve the most similar sequences of S in $\mathcal{S}$.

Other examples include (multiple) sequence alignment, primer design, literature search, phylogenetic analysis, etc.

7.4 Sequence Alignment

Alignment is frequently used to identify regions of similarity between sequences. In biology, significant similarity can be a consequence of functional, structural, or evolutionary relationships between the sequences. Alignment can be considered for two sequences or more sequences.

An *alignment* between two sequences $X = s_1...s_m$ and $Y = t_1...t_n$ is a mapping[2] between positions in the two sequences. In addition to the normal

[1] Some of these problems were discussed in [117, 103].

[2] The mapping must satisfy certain properties to ensure that the aligned sequences can be displayed in the manner illustrated in Figure 7.1.

positions in the sequences, gaps (represented by -) are allowed. A gap indicates an insertion in one sequence or a deletion in the other sequence. In the aligned sequences, if a column contains -, then the column is an *indel*; if a column does not contain - and it contains the same element in both rows, then the column is a *match*; otherwise the column is a *mismatch*.

```
ACTCCTC-A
AG-CC-CCA
```

Fig. 7.1. An alignment between two sequences

Figure 7.1 shows an alignment between $X = ACTCCTCA$ and $Y = AGCCCCA$. In the figure, column 1 is a match, column 2 is a mismatch, and column 3 is an indel; there are a total of 5 matches, 1 mismatch, and 3 indels.

The quality of an alignment is measured by three numbers: the number of matches, the number of mismatches, and the number of indels. The three numbers can be combined into a formula to define the quality of the alignment, depending on the application.

In some applications, different mismatches are viewed differently. An element x can be more similar to an element y than to an element z. Similarity between elements can be given by a matrix. The objective is then to find alignments which optimize the aggregated similarity of the elements in the matching columns.

Alignment of three or more sequences is similar, and is called *multiple alignment*. Figure 7.2 gives an example.

```
VLRQAAQ--QVLQRQIIQGPQQ
VLRQVVQ--QALQRQIIQGPQQ
VLRQAAHLAQQLYQGQ----RQ
VLRQAAH--QQLYQGQ----RQ
```

Fig. 7.2. A multiple alignment of sequences

Given a set of sequences there can be many possible alignments. Among these, there is an optimal alignment, for a given quality measure. The optimal alignment can be computed using dynamic programming. Sometimes speed of computation is important. In such cases, heuristic methods can be used. For example, one can first find possible perfect short matches, and then extend or join these perfect short matches to find good long matches. Sequence alignment is a thoroughly studied problem. Interested readers can find more details in, for example, [40, 25].

7.5 Biological Sequence Databases and Biological Data Analysis Resources

Many biological databases and resources are available at the National Center for Biotechnology Information (NCBI), which was established in 1988 as a national resource for molecular biology information. The list of data and analysis resources provided by NCBI is long, including analysis and retrieval resources for the data in GenBank. More information can be found at the website `www.ncbi.nlm.nih.gov` as well as from [121].

Protein sequences are an especially interesting category of biological sequences since protein is functionally essential in life and its alphabet is large (20 amino acids). There are several well-known protein databases: Pfam [7] is a collection of protein families and domains. NCBI also provides retrieval of protein sequence data. Swiss-Prot [6] is a protein sequence database which strives to provide a high level of annotation, a minimal level of redundancy and high level of integration with other databases.

Alternative splicing is widespread in mammalian gene expression, and variant splice patterns are often specific to different stages of development, particular tissues or a disease state. ASD [107] is a database of computationally delineated alternative splice events as seen in alignments of EST/cDNA sequences with genome sequences, and a database of alternatively spliced exons collected from literature. ASD is available at `http://www.ebi.ac.uk/asd`.

The International HapMap Project is a multi-country effort to identify and catalog genetic similarities and differences in human beings. The data collected by this project are certain subsequences of the human DNA sequences, for sequence positions where the DNA can be different among different individuals. The data can help researchers to find genes that affect health, disease, and individual responses to medications and environmental factors. More information about the project as well as the data collected from this project can be found at `http://www.hapmap.org/` and from [36].

References

1. R. Agrawal, D. Gunopulos, and F. Leymann. Mining process models from workflow logs. In *Proceedings of the 6th International Conference on Extending Database Technology (EDBT)*, pages 469–483, London, UK, 1998. Springer-Verlag.
2. R. Agrawal, T. Imielinski, and A. Swami. Mining association rules between sets of items in large databases. In *Proc. ACM-SIGMOD Int. Conf. Management of Data (SIGMOD)*, pages 207–216, Washington, DC, May 1993.
3. R. Agrawal and R. Srikant. Mining sequential patterns. In *Proc. Int. Conf. Data Engineering (ICDE)*, pages 3–14, Taipei, Taiwan, Mar. 1995.
4. S. Altschul et al. Gapped BLAST and PSI-BLAST: a new generation of protein database search programs. *Nucleic Acids Research*, 25(17):3389–3402, 1997.
5. J. Ayres, J. Flannick, J. Gehrke, and T. Yiu. Sequential pattern mining using a bitmap representation. In *Proc. ACM SIGKDD Int. Conf. on Knowledge Discovery and Data Mining (KDD)*, pages 429–435, Edmonton, Alberta, Canada, July 2002.
6. A. Bairoch and R. Apweiler. The SWISS-PROT protein sequence database and its supplement TrEMBL in 2000. *Nucleic Acids Research*, 28(1):45–48, 2000.
7. A. Bateman, E. Birney, L. Cerruti, R. Durbin, L. Etwiller, S. R. Eddy, S. Griffiths-Jones, K. L. Howe, M. Marshall, and E. L. L. Sonnhammer. The Pfam Protein Families Database. *Nucleic Acids Research*, 30(1):276–280, 2002.
8. A. Ben-Dor, B. Chor, R. Karp, and Z. Yakhini. Discovering local structure in gene expression data: the order-preserving submatrix problem. In *Proceedings of the sixth annual international conference on Computational biology*, pages 49–57, Washington, DC, USA, 2002. ACM Press.
9. D. J. Berndt and J. Clifford. Finding patterns in time series: a dynamic programming approach. *Advances in knowledge discovery and data mining*, pages 229–248, 1996.
10. G. E. P. Box and G. Jenkins. *Time Series Analysis, Forecasting and Control.* Holden-Day, Incorporated, 1990.
11. Graham Brightwell and Peter Winkler. Counting linear extensions is #p-complete. In *Proceedings of the twenty-third annual ACM symposium on Theory of computing*, pages 175–181, New Orleans, Louisiana, United States, 1991. ACM Press.

12. C. J. C. Burges. A Tutorial on Support Vector Machines for Pattern Recognition. *Data Mining and Knowledge Discovery*, 2(2):121–167, 1998.
13. J. Burke, D. Davison, and W. Hide. d2_cluster: A Validated Method for Clustering EST and Full-Length cDNA Sequences. *Genome Research*, pages 1135–1142, 1999.
14. H. Cao, D. W. Cheung, and N. Mamoulis. Discovering Partial Periodic Patterns in Discrete Data Sequences. *Proceedings of The 8th Pacific-Asia Conference on Knowledge Discovery and Data Mining*, 2004.
15. G. Casella and E. I. George. Explaining the Gibbs sampler. *The American statistician*, 46(3):167–174, 1992.
16. L. Chen and G. Dong. Succinct and informative cluster descriptions for document repositories. In *Proceedings of International Conference on Web-Age Information Management*, pages 109–121, 2006.
17. Y. Chen, G. Dong, J. Han, B. W. Wah, and J. Wang. Multi-dimensional regression analysis of time-series data streams. *Proc. VLDB*, pages 323–334, 2002.
18. Y. Chi, Y. Yang, and R. R. Muntz. HybridTreeMiner: An Efficient Algorithm for Mining Frequent Rooted Trees and Free Trees Using Canonical Forms. *Proceedings of the 16th International Conference on Scientific and Statistical Database Management (SSDBM)*, 2004.
19. D. Y. Chiu, Y. H. Wu, and A. L. P. Chen. An efficient algorithm for mining frequent sequences by a new strategy without support counting. In *Proceedings of the twentieth IEEE International Conference on Data Engineering (ICDE'04)*, pages 275–286, Boston, Massachusetts, United States, 2004. IEEE Computer Society.
20. K. C. Chou. Prediction of protein cellular attributes using pseudo-amino acid composition. *Proteins Structure Function and Genetics*, 44(1):60–60, 2001.
21. D. J. Cook and L. B. Holder. Graph-based data mining. *IEEE Intelligent Systems*, 15(2):32–41, 2000.
22. D. Daniels, P. Zuber, and R. Losick. Two amino acids in an RNA polymerase sigma factor involved in the recognition of adjacent base pairs in the-10 region of a cognate promoter. *Proc Natl Acad Sci US A*, 87(20):8075–8079, 1990.
23. M. O. Dayhoff, R. M. Schwartz, and B. C. Orcutt. A model of evolutionary change in proteins. *Atlas of Protein Sequence and Structure*, 5(Suppl 3):345–352, 1978.
24. A. L. Delcher, D. Harmon, S. Kasif, O. White, and S. L. Salzberg. Improved microbial gene identification with GLIMMER. *Nucleic Acids Research*, 27(23):4636–4641, 1999.
25. R. Durbin, A. Krogh, G. Mitchison, and S. R. Eddy. *Biological Sequence Analysis: Probabilistic Models of Proteins and Nucleic Acids.* Cambridge University Press, 1998.
26. R. C. Edgar. A comparison of scoring functions for protein sequence profile alignment. *Bioinformatics*, 20(8):1301–1308, 2004.
27. M. G. Elfeky, W. G. Aref, and A. K. Elmagarmid. Incremental, Online, and Merge Mining of Partial Periodic Patterns in Time-Series Databases. *IEEE Transactions on Knowledge and Data Engineering*, 16(3):332–342, 2004.
28. M. G. Elfeky, W. G. Aref, and A. K. Elmagarmid. Using Convolution to Mine Obscure Periodic Patterns in One Pass. *Proceedings of the 9th International Conference on Extending Database Technology (EDBT)*, pages 605–620, 2004.

29. M. G. Elfeky, W. G. Aref, and A. K. Elmagarmid. Periodicity Detection in Time Series Databases. *IEEE Transactions on Knowledge and Data Engineering*, 17(7):875–887, 2005.
30. A. J. Enright and C. A. Ouzounis. GeneRAGE: a robust algorithm for sequence clustering and domain detection. *Bioinformatics*, 16(5):451–457, 2000.
31. A. J. Enright, S. Van Dongen, and C. A. Ouzounis. An efficient algorithm for large-scale detection of protein families. *Nucleic Acids Research*, 30(7):1575–1584, 2002.
32. D. H. Fisher. Knowledge acquisition via incremental conceptual clustering. *Machine Learning*, 2(2):139–172, 1987.
33. M. Garey and D. Johnson. *Computers and Intractability: a Guide to The Theory of NP-Completeness.* Freeman and Company, New York, 1979.
34. M. Garofalakis, R. Rastogi, and K. Shim. SPIRIT: Sequential pattern mining with regular expression constraints. In *Proc. Int. Conf. Very Large Data Bases (VLDB)*, pages 223–234, Edinburgh, UK, Sept. 1999.
35. P. Geurts, A. B. Cuesta, and L. Wehenkel. Segment and Combine Approach for Biological Sequence Classification. *Proceedings of IEEE Symposium on Computational Intelligence in Bioinformatics and Computational Biology (CIBCB'05)*, pages 1–8, 2005.
36. R. A. Gibbs, J. W. Belmont, P. Hardenbol, T. D. Willis, F. Yu, H. Yang, L. Y. Ch'ang, W. Huang, B. Liu, Y. Shen, et al. The International HapMap Project. *Nature*, 426(6968):789–796, 2003.
37. A. Gionis, T. Kujala, and H. Mannila. Fragments of order. In *Proceedings of the ninth ACM SIGKDD international conference on Knowledge discovery and data mining*, pages 129–136. ACM Press, 2003.
38. W. N. Grundy, T. L. Bailey, C. Elkan, and M. E. Baker. Meta-MEME: motif-based hidden Markov models of protein families. *Computer Applications in the Biosciences*, 13(4):397–406, 1997.
39. J. Guo, Y. Lin, and Z. Sun. A novel method for protein subcellular localization: Combining residue-couple model and SVM. *Proceedings of the 3rd Asia-Pacific Bioinformatics Conference*, pages 117–129, 2005.
40. D. Gusfield. *Algorithms on Strings, Trees and Sequences: Computer Science and Computational Biology.* Cambridge University Pr., 1997.
41. R. Gwadera, M. J. Atallah, and W. Szpankowski. Reliable detection of episodes in event sequences. *Knowledge and Information Systems*, 7(4):415–437, 2005.
42. J. Han, G. Dong, and Y. Yin. Efficient mining of partial periodic patterns in time series database. *Proceedings of IEEE International Conference on Data Mining*, pages 106–115, 1999.
43. J. Han, W. Gong, and Y. Yin. Mining segment-wise periodic patterns in time-related databases. *Proc. Int. Conf. on Knowledge Discovery and Data Mining*, pages 214–218, 1998.
44. J. Han and M. Kamber. *Data Mining: Concepts and Techniques.* Morgan Kaufmann, 2006.
45. J. Han, J. Pei, B. Mortazavi-Asl, Q. Chen, U. Dayal, and M.-C. Hsu. FreeSpan: Frequent pattern-projected sequential pattern mining. In *Proc. ACM SIGKDD Int. Conf. Knowledge Discovery in Databases (KDD)*, pages 355–359, Boston, MA, Aug. 2000.
46. D. J. Hand, H. Mannila, and P. Smyth. *Principles of Data Mining.* Bradford Books, 2001.

47. E. Hartuv and R. Shamir. Clustering algorithm based on graph connectivity. *INF. PROCESS. LETT*, 76(4):175–181, 2000.
48. M. A. Hearst. Untangling text data mining. *Proceedings of ACL*, 99:20–26, 1999.
49. S. Henikoff and J. G. Henikoff. Performance evaluation of amino acid substitution matrices. *Proteins: Structure, Function, and Genetics*, 17:49–61, 1993.
50. L. Holm and C. Sander. Protein structure comparison by alignment of distance matrices. *J. Mol. Biol*, 233(1):123–128, 1993.
51. S. H. Huang, R. S. Liu, C. Y. Chen, Y. T. Chao, and S. Y. Chen. Prediction of Outer Membrane Proteins by Support Vector Machines Using Combinations of Gapped Amino Acid Pair Compositions. *Proceedings of the Fifth IEEE Symposium on Bioinformatics and Bioengineering*, pages 113–120, 2005.
52. G. Hulten, L. Spencer, and P. Domingos. Mining time-changing data streams. *Proceedings of the seventh ACM SIGKDD international conference on Knowledge discovery and data mining*, pages 97–106, 2001.
53. A. Inokuchi, T. Washio, and H. Motoda. An apriori-based algorithm for mining frequent substructures from graph data. In *Proc. European Symp. Principle of Data Mining and Knowledge Discovery (PKDD)*, pages 13–23, Lyon, France, Sept. 2000.
54. A. K. Jain, J. Mao, and K. M. Mohiuddin. Artificial Neural Networks: A Tutorial. *Computer*, 29(3):31–44, 1996.
55. X. Ji, J. Bailey, and G. Dong. Mining Minimal Distinguishing Subsequence Patterns with Gap Constraints. *Proceedings of the Fifth IEEE International Conference on Data Mining*, pages 194–201, 2005.
56. X. Ji, J. Bailey, and G. Dong. Mining Minimal Distinguishing Subsequence Patterns with Gap Constraints. *Knowledge and Information Systems*, 11(3):259–296, 2007.
57. S. C. Johnson. Hierarchical clustering schemes. *Psychometrika*, 32(3):241–254, 1967.
58. D. T. Jones. Protein secondary structure prediction based on position-specific scoring matrices. *J. Mol. Biol*, 292(2):195–202, 1999.
59. K. Karplus, C. Barrett, and R. Hughey. Hidden Markov models for detecting remote protein homologies. *Bioinformatics*, 14:846–856, 1998.
60. H. Kawaji, Y. Yamaguchi, H. Matsuda, and A. Hashimoto. A graph-based clustering method for a large set of sequences using a graph partitioning algorithm. *Genome Informatics*, 12:93–102, 2001.
61. E. Keogh and S. Kasetty. On the Need for Time Series Data Mining Benchmarks: A Survey and Empirical Demonstration. *Data Mining and Knowledge Discovery*, 7(4):349–371, 2003.
62. A. Krogh, M. Brown, IS Mian, K. Sjolander, and D. Haussler. Hidden Markov models in computational biology. Applications to protein modeling. *J Mol Biol*, 235(5):1501–31, 1994.
63. C. E. Lawrence, S. F. Altschul, M. S. Boguski, J. S. Liu, A. F. Neuwald, and J. C. Wootton. Detecting subtle sequence signals: a Gibbs sampling strategy for multiple alignment. *Science*, 262(5131):208, 1993.
64. M. Li, J. H. Badger, X. Chen, S. Kwong, P. Kearney, and H. Zhang. An information-based sequence distance and its application to whole mitochondrial genome phylogeny. *Bioinformatics*, 17(2):149–154, 2001.

65. Y. Li, P. Ning, X. S. Wang, and S. Jajodia. Discovering calendar-based temporal association rules. *Proceedings of International Symposium on Temporal Representation and Reasoning*, pages 111–118, 2001.
66. B. Liu, W. Hsu, and Y. Ma. Mining association rules with multiple minimum supports. *Proceedings of the fifth ACM SIGKDD international conference on Knowledge discovery and data mining*, pages 337–341, 1999.
67. H. Liu and H. Motoda. *Feature Selection for Knowledge Discovery and Data Mining.* Springer, 1998.
68. J. Liu and W. Wang. Op-cluster: Clustering by tendency in high dimensional space. In *Proceedings of the Third IEEE International Conference on Data Mining (ICDM)*, Melbourne, Florida, Nov. 2003. IEEE.
69. X. Liu, D. L. Brutlag, and J. S. Liu. BioProspector: discovering conserved DNA motifs in upstream regulatory regions of co-expressed genes. *Pac. Symp. Biocomput*, 6:127–138, 2001.
70. H. J. Loether and D. G. McTavish. *Descriptive and Inferential Statistics: An Introduction.* Allyn and Bacon, 1993.
71. A. V. Lukashin and M. Borodovsky. GeneMark.hmm: new solutions for gene finding. *Nucleic Acids Research*, 26(4):1107–1115, 1998.
72. Q. Ma, J. T. L. Wang, D. Shasha, and C. H. Wu. DNA sequence classification via an expectation maximizationalgorithm and neural networks: a case study. *IEEE Transactions on Systems, Man and Cybernetics, Part C*, 31(4):468–475, 2001.
73. S. Ma and J. L. Hellerstein. Mining partially periodic event patterns with unknown periods. *Proceedings of IEEE International Conference on Data Engineering*, pages 205–214, 2001.
74. H. Mannila and C. Meek. Global partial orders from sequential data. In *Proc. ACM SIGKDD Int. Conf. Knowledge Discovery in Databases (KDD)*, pages 150–160, Boston, MA, Aug. 2000.
75. H. Mannila, H. Toivonen, and A. I. Verkamo. Discovering frequent episodes in sequences. In *Proc. Int. Conf. Knowledge Discovery and Data Mining (KDD)*, pages 210–215, Montreal, Canada, Aug. 1995.
76. H. Mannila, H. Toivonen, and A. I. Verkamo. Discovery of frequent episodes in event sequences. *Data Mining and Knowledge Discovery*, 1:259–289, 1997.
77. F. Masseglia, F. Cathala, and P. Poncelet. The PSP approach for mining sequential patterns. In *Proc. European Symp. Principle of Data Mining and Knowledge Discovery (PKDD)*, pages 176–184, Nantes, France, Sept. 1998.
78. L. A. McCue, W. Thompson, C. S. Carmack, and C. E. Lawrence. Factors Influencing the Identification of Transcription Factor Binding Sites by Cross-Species Comparison. *Genome Research*, 2002.
79. R. S. Michalski and R. E. Stepp. Learning from observation: Conceptual clustering. *Machine Learning: An Artificial Intelligence Approach*, 1:331–363, 1983.
80. S. Mika, G. Ratsch, J. Weston, B. Scholkopf, and K. R. Mullers. Fisher discriminant analysis with kernels. *Neural Networks for Signal Processing IX, Proceedings of IEEE Signal Processing Society Workshop*, pages 41–48, 1999.
81. R. Ng, L. V. S. Lakshmanan, J. Han, and A. Pang. Exploratory mining and pruning optimizations of constrained associations rules. In *Proc. ACM-SIGMOD Int. Conf. Management of Data (SIGMOD)*, pages 13–24, Seattle, WA, June 1998.

82. B. Ozden, S. Ramaswamy, and A. Silberschatz. Cyclic association rules. *Proceedings of IEEE International Conference on Data Engineering*, pages 412–421, 1998.
83. K. J. Park and M. Kanehisa. Prediction of protein subcellular locations by support vector machines using compositions of amino acids and amino acid pairs. *Bioinformatics*, 19(13):1656–1663, 2003.
84. N. Pasquier, Y. Bastide, R. Taouil, and L. Lakhal. Discovering frequent closed itemsets for association rules. In *Proc. 7th Int. Conf. Database Theory (ICDT)*, pages 398–416, Jerusalem, Israel, Jan. 1999.
85. W. R. Pearson. Rapid and sensitive sequence comparison with FASTP and FASTA. *Methods Enzymol*, 183:63–98, 1990.
86. J. Pei and J. Han. Can we push more constraints into frequent pattern mining? In *Proc. ACM SIGKDD Int. Conf. Knowledge Discovery in Databases (KDD)*, pages 350–354, Boston, MA, Aug. 2000.
87. J. Pei, J. Han, and L. V. S. Lakshmanan. Mining frequent itemsets with convertible constraints. In *Proc. Int. Conf. Data Engineering (ICDE)*, pages 433–332, Heidelberg, Germany, April 2001.
88. J. Pei, J. Han, B. Mortazavi-Asl, H. Pinto, Q. Chen, U. Dayal, and M.-C. Hsu. PrefixSpan: Mining sequential patterns efficiently by prefix-projected pattern growth. In *Proc. Int. Conf. Data Engineering (ICDE)*, pages 215–224, Heidelberg, Germany, April 2001.
89. J. Pei, J. Han, and W. Wang. Constraint-based sequential pattern mining in large databases. In *Proc. Int. Conf. on Information and Knowledge Management (CIKM)*, McLean, VA, Nov. 2002.
90. D. S. Prestridge. Predicting Pol II promoter sequences using transcription factor binding sites. *J Mol Biol*, 249(5):923–32, 1995.
91. D. Pribnow. Nucleotide Sequence of an RNA Polymerase Binding Site at an Early T7 Promoter. *Proceedings of the National Academy of Sciences*, 72(3):784–788, 1975.
92. L. R. Rabiner. A tutorial on hidden Markov models and selected applications inspeech recognition. *Proceedings of the IEEE*, 77(2):257–286, 1989.
93. J. F. Roddick and M. Spiliopoulou. A survey of temporal knowledge discovery paradigms and methods. *IEEE Transactions on Knowledge and Data Engineering*, 14(4):750–767, 2002.
94. T. D. Schneider and R. M. Stephens. Sequence logos: a new way to display consensus sequences. *Nucleic Acids Res*, 18(20):6097–6100, 1990.
95. R. She, F. Chen, K. Wang, M. Ester, J. L. Gardy, and F. S. L. Brinkman. Frequent-Subsequence-Based Prediction of Outer Membrane Proteins. In *Proceedings of KDD*, 2003.
96. J. Simon. On the difference between the one and the many. *Automata, Languages, and Programming, Lecture Notes in Computer Science*, 52:480–491, 1977.
97. C. W. Smith and J. Valcarcel. Alternative pre-mRNA splicing: the logic of combinatorial control. *Trends Biochem. Sci*, 25(8):381–388, 2000.
98. T. F. Smith and M. S. Waterman. Identification of Common Molecular Subsequences. *J. Mol. Bwl*, 147:195–197, 1981.
99. P. Smyth. Clustering sequences with hidden Markov models. *Advances in Neural Information Processing Systems*, 9:648–654, 1997.
100. S. Sonnenburg, G. Ratsch, and C. Schafer. Learning interpretable SVMs for biological sequence classification. *RECOMB, LNBI*, 3500:389–407, 2005.

101. R. Srikant and R. Agrawal. Mining sequential patterns: Generalizations and performance improvements. In *Proc. 5th Int. Conf. Extending Database Technology (EDBT)*, pages 3–17, Avignon, France, Mar. 1996.
102. M. Steinbach, G. Karypis, and V. Kumar. A comparison of document clustering techniques. *KDD Workshop on Text Mining*, 34:35, 2000.
103. R. Stevens, C. Goble, P. Baker, and A. Brass. A classification of tasks in bioinformatics. *Bioinformatics*, 17(2):180–188, 2001.
104. G. D. Stormo. DNA binding sites: representation and discovery. *Bioinformatics*, 16(1):16–23, 2000.
105. P. N. Tan, M. Steibach, and Kumar V. *Introduction to Data Mining.* Addison-Wesley, 2005.
106. C. Tang, R. W. H. Lau, H. Yin, Q. Li, Y. Lu, Z. Yu, L. Xiang, and T. Zhang. Discovering Tendency Association between Objects with Relaxed Periodicity and its Application in Seismology. *Proceedings of ICSC, LNCS Vol. 1749, 51*, 62, 1999.
107. T. A. Thanaraj, S. Stamm, F. Clark, J. J. Riethoven, V. Le Texier, and J. Muilu. ASD: the Alternative Splicing Database. *Nucleic Acids Research*, 32(90001):W181–W186, 2004.
108. D. C. Torney, C. Burks, D. Davison, and K. M. Sirotkin. Computation of d2: A Measure of Sequence Dissimilarity. *Computers and DNA*, pages 109–125, 1990.
109. P. Tzvetkov, X. Yan, and J. Han. TSP: Mining top-k closed sequential patterns. In *Proceedings of the Third IEEE International Conference on Data Mining (ICDM)*, Melbourne, Florida, Nov. 2003. IEEE.
110. J. Valdes, R. E. Tarjan, and E. L. Lawler. The recognition of series parallel digraphs. In *Proceedings of the eleventh annual ACM symposium on Theory of computing*, pages 1–12, Atlanta, Georgia, United States, 1979. ACM Press.
111. L. G. Valiant. The complexity of computing the permanent. *Theoretical Computer Science*, 8:189–201, 1977.
112. W. van der Aalst, T. Weijters, and L. Maruster. Workflow mining: Discovering process models from event logs. *IEEE Transactions on Knowledge and Data Engineering*, 16:1128–1142, September 2004.
113. V. N. Vapnik. *Statistical learning theory.* Wiley, 1998.
114. J. Wang and J. Han. BIDE: Efficient mining of frequent closed sequences. In *Proceedings of the twentieth IEEE International Conference on Data Engineering*, pages 79–90, Boston, Massachusetts, United States, 2004. IEEE Computer Society.
115. J. Wang, J. Han, and J. Pei. CLOSET+: Searching for the best strategies for mining frequent closed itemsets. In *Proceedings of the Nineth ACM SIGKDD International Conference on Knowledge Discovery and Data Mining (KDD'03)*, Washington, D.C, 2003. ACM Press.
116. J. T. L. Wang, Q. Ma, D. Shasha, and C. H. Wu. New techniques for extracting features from protein sequences. *IBM Systems Journal*, 40(2):426–441, 2001.
117. J. T. L. Wang, M. J. Zaki, H. T. T. Toivonen, and D. Shasha, editors. *Data Mining in Bioinformatics.* Springer, 2005.
118. W. Wang and O. R. Zaiane. Clustering Web sessions by sequence alignment. *Proceedings of 13th International Workshop on Database and Expert Systems Applications*, pages 394–398, 2002.
119. T. Washio and H. Motoda. State of the art of graph-based data mining. *ACM SIGKDD Explorations Newsletter*, 5(1):59–68, 2003.

120. G. M. Weiss. Mining with rarity: a unifying framework. *ACM SIGKDD Explorations Newsletter*, 6(1):7–19, 2004.
121. D. L. Wheeler, T. Barrett, D. A. Benson, S. H. Bryant, K. Canese, V. Chetvernin, D. M. Church, M. DiCuccio, R. Edgar, S. Federhen, et al. Database resources of the National Center for Biotechnology Information. *Nucleic Acids Research*, 35(Database issue):D5, 2007.
122. W. J. Wilbur. On the PAM matrix model of protein evolution. *Molecular Biology and Evolution*, 2:434–447, 1985.
123. C. Wu, M. Berry, S. Shivakumar, and J. McLarty. Neural networks for full-scale protein sequence classification: Sequence encoding with singular value decomposition. *Machine Learning*, 21(1):177–193, 1995.
124. X. Yan and J. Han. CloseGraph: Mining closed frequent graph patterns. *Proc. of the 9th ACM SIGKDD International Conference on Knowledge Discovery and Data Mining*, pages 286–295, 2003.
125. X. Yan, J. Han, and R. Afshar. CloSpan: Mining closed sequential patterns in large databases. In *Proc. SIAM Int. Conf. Data Mining*, San Fransisco, CA, May 2003.
126. J. Yang and W. Wang. CLUSEQ: efficient and effective sequence clustering. *Proceedings. 19th International Conference on Data Engineering*, pages 101–112, 2003.
127. J. Yang, W. Wang, and P. S. Yu. Mining Asynchronous Periodic Patterns in Time Series Data. *IEEE Transactions on Knowledge and Data Engineering*, 15(3):613–628, 2003.
128. J. Yang, W. Wang, and P. S. YU. Discovering High-Order Periodic Patterns. *Knowledge and Information Systems*, 6(3):243–268, 2004.
129. J. Yang, W. Wang, and P. S. Yu. Mining Surprising Periodic Patterns. *Data Mining and Knowledge Discovery*, 9(2):189–216, 2004.
130. M. Zaki. Generating non-redundant association rules. In *Proc. ACM SIGKDD Int. Conf. Knowledge Discovery in Databases (KDD)*, pages 34–43, Boston, MA, Aug. 2000.
131. M. Zaki. Efficiently mining frequent trees in a forest. *Proceedings of the eighth ACM SIGKDD international conference on Knowledge discovery and data mining*, pages 71–80, 2002.
132. M. J. Zaki. SPADE: An efficient algorithm for mining frequent sequences. *Mach. Learn.*, 42(1-2):31–60, 2001.
133. M. J. Zaki and C. J. Hsiao. CHARM: An efficient algorithm for closed itemset mining. In *Proc. SIAM Int. Conf. Data Mining*, pages 457–473, Arlington, VA, April 2002.
134. C. T. Zhang and J. Wang. Recognition of protein coding genes in the yeast genome at better than 95% accuracy based on the Z curve. *Nucleic Acids Research*, 28(14):2804–2814, 2000.

Index